Mass-Observation and Everyday Life

Mass-Observation and Everyday Life

Culture, History, Theory

Nick Hubble

First published 2006 by
PALGRAVE MACMILLAN
Houndmills, Basingstoke, Hampshire RG21 6XS and
175 Fifth Avenue, New York, N. Y. 10010
Companies and representatives throughout the world

PALGRAVE MACMILLAN is the global academic imprint of the Palgrave Macmillan division of St. Martin's Press, LLC and of Palgrave Macmillan Ltd. Macmillan® is a registered trademark in the United States, United Kingdom and other countries. Palgrave is a registered trademark in the European Union and other countries.

ISBN-13: 978–1–4039–3555–7 hardback
ISBN-10: 1–4039–3555–6   hardback

This book is printed on paper suitable for recycling and made from fully managed and sustained forest sources.

A catalogue record for this book is available from the British Library.

Library of Congress Cataloging-in-Publication Data
Hubble, Nick, 1965–
    Mass-Observation and everyday life : culture, history, theory / Nick Hubble.
        p. cm.
    Includes bibliographical references and index.
    ISBN 1–4039–3555–6 (cloth)
        1. Mass-Observation. 2. Social history–20th century. 3. Social history–Historiography. 4. Social surveys–Great Britain–History–20th century. 5. Great Britain–Social conditions–20th century. I. Title.
HN16.H83 2005   2005048470
306′.0941′0904–dc22

10   9   8   7   6   5   4   3   2   1
15 14 13 12 11 10 09 08 07 06

Transferred to Digital Printing 2007

For Andrea Hammel

Contents

List of Abbreviations

ARP	Air Raid Precautions
ATS	Auxiliary Territorial Service
BBC	British Broadcasting Corporation
BIPO	British Institute of Public Opinion
CMP	Charles Madge Papers
CP	Communist Party
EMB	Empire Marketing Board
FR	File Report
GPO	General Post Office
GSS	Government Social Survey
HI	Home Intelligence
LSE	London School of Economics
MAP	Mary Adams Papers
M-O	Mass-Observation
M-O A	Mass-Observation Archive
MOI	Ministry of Information
MP	Member of Parliament
NIESR	National Institute of Economic and Social Research
NUWM	National Unemployed Workers Movement
PEP	Political and Economic Planning
PP	Popular Poetry
RIO	Regional Intelligence Officer
WSS	Wartime Social Survey

Acknowledgements

This book would not have been possible without the support of my partner, Andrea Hammel, and our children Sara and Max. Special thanks are also due to the help over the years from Christine and Brian Hubble and Marion and Karl Ludwig Hammel.

I am grateful to the organisers of the following conferences for allowing me to develop my ideas on Mass-Observation: 'Surveillance: An Interdisciplinary Conference', Liverpool JMU, June 1998; 'The Literature of the 1930s', Anglia Polytechnic University, November 2000; 'Englishness in all its Diversity', University of Central England, July 2003; and 'The Noise of History: Literature and Culture in the 1930s', Trinity College Carmarthen and the Dylan Thomas Centre, Swansea, November 2003. I have benefited immeasurably from giving seminar papers to the Centre for Life History Research Seminar Series, University of Sussex, in November 1998 and again in March 2004; the English Graduate Colloquium, University of Sussex, January 1999; the English Research Seminar, University of Central England, October 2003; and the Centre for Urban History, University of Leicester, February 2004. I have also found the contemporary academic experience of interview presentations to be a particularly fertile ground for rigorous feedback.

Drew Milne helped launch this project as a DPhil thesis and Tony Inglis provided indispensable help when he took over its supervision. I hope this (first) instalment of the finished product will provide some recompense to them for their time and commitment. Equal thanks are due to the continuing support of Dorothy Sheridan and the staff at the Mass-Observation Archive including, over the years, Fiona Courage, Joy Eldridge, Anna Green, Simon Homer, Sandra Koa Wing, Helen Monk, Lucy Pearson, Judy Pickering and Karen Watson. Indeed, all the staff at the University of Sussex Library have been helpful and friendly over the past decade during which I have been a user. I am particularly grateful to Dorothy and the Trustees of the Archive for the Honorary Research Fellowship they granted me for the period of writing this book.

Thanks also to the following: Melanie Blair, Libby and Dan Danahar, Ben Highmore, Tristram Hooley, Sharon Krummel, Jill Lake, Margaretta Jolly, Laura Marcus, Niamh Moore, Francis Mulhern, Jen Nelson, Briar

Towers, all my friends and former colleagues in the English department at the University of Central England and all my new friends and colleagues in the Centre for Suburban Studies and the School of Humanities at Kingston University.

I would also like to thank the Trustees of the Mass-Observation Archive for permission to quote Mass-Observation material and Vicky Randall for permission to quote from the papers of her father, Charles Madge.

Introduction: The Mass-Observation Project

Defining Mass-Observation

M-O has always been an enigma since its foundation in 1937. Defini-
tions shift bewilderingly according to perspective: 'It has been characterised variously as a documentary or photographic project (Laing
1980), as a deeply flawed social survey (Abrams 1951), as a middle-class
adventure at the expense of the working class (Gurney 1997), as salvationist (Hynes [1982]), as a people's history (Calder 1985), and as a life
history project which was a precursor to, for example, present-day oral
history (Sheridan 1996)' (Sheridan et al 2000: 27). This suggests that a
multidisciplinary and nuanced approach is required to do full justice to
its complexity. However, such approaches often struggle to maintain
focus and can become subject to historicist pressures that reduce complexity to the product of a particular time and place. For instance,
Samuel Hynes's description of M-O in his influential *The Auden Generation* establishes it as a paradigmatic example of a radical generation's
concerns while simultaneously reducing it to the background of social
context against which his principle literary subjects are foregrounded:
'It was at once literary and scientific, realist and surrealist, political and
psychological, Marxist and Freudian, objective and salvationist. In its
confusions of methods and goals it is a complex example of the confusions of young intellectuals at the time' (Hynes 1982: 278). One way of
evading these historicist pitfalls is to pursue potentially open-ended
lines of enquiry as exemplified by the title of Penny Summerfield's essay
'Mass-Observation: Social Research or Social Movement?' (Summerfield
1985). Yet investigations of M-O as a movement have tended to founder
on the question of what its goals were. One brief, but nonetheless
intriguing, discussion is to be found in the excellent M-O anthology

with historical commentary, *Speak for Yourself*, which acknowledges the significance of M-O's close links to Surrealist and documentary movements: 'In both cases, one aim was a kind of social therapy; in either case, a further aim might be social transformation' (Calder and Sheridan 1984: 4). These two approaches in their counterpart Continental forms of the French *Le Surréalisme au service de la Révolution* and the German *die neue Sachlichkeit* occupied prominent and rival positions in the European Left's debates concerning aesthetics and politics in the 1930s, as will be discussed in chapter three. The stakes involved in determining what kind of social therapy and social transformation were required rose to impossible levels with the seemingly inexorable advance of the Right as the decade progressed, informing Walter Benjamin's statement: '[Humanity's] self-alienation has reached such a degree that it can experience its own destruction as an aesthetic pleasure of the first order. This is the situation of politics which Fascism is rendering aesthetic. Communism responds by politicising art' (Benjamin 1999: 235)

All the above contradictions and ambiguities are highlighted by the choice of name which from the very beginning prompted critics to question 'whether mass-observation means observation *of* the mass or observation *by* the mass' (Marshall 1937: 49; see also Calder 1986: xiii). The working-class writer Jack Common's quip about 'the attempts of nice young men to penetrate into working-class pubs and try to get to know the workers' might have been a broad jibe at the wider documentary movement and its public-school-educated, popular-front-supporting and sometimes homosexual writers and photographers, but his real complaint was that the shared approach behind both social-realist fiction and the social anthropology adopted by M-O could only ever define the masses negatively in contrast to middle-class individualism: 'Even in supposedly sympathetic works of fiction the man of the masses registers as a thwarted individualist whose tragedy is, there's a lot of things he hasn't got, and he ought to have them because he has much amiable virtue – I'd hate to have anyone draw such a picture of me. So would you, I'll bet' (Common 1988: 2–3). This theme was later to become central to the seminal text of British Cultural Studies: Raymond Williams's *Culture and Society*. In his conclusion to that book – divided under a number of sub-headings including 'Mass and Masses', 'Mass-communication' and 'Mass-observation' – Williams famously stated: 'There are in fact no masses; there are only ways of seeing people as masses' (Williams 1990: 300). This assertion gains particular political resonance when directed at M-O because of their investigation of wartime public morale for the MOI.

A closely related issue is whether the opinion gathering methods employed by M-O – in many ways the direct precursors of the sampling and focus group work that has become so influential in Britain since the 1990s – actually served to transform rather than reflect public opinion during the Second World War. On the other hand, as Stuart Laing has shown, it is possible to identify positive connotations for 'mass' as used by M-O. Among the implications he identifies in their work are the contention 'that a new kind of society had arisen in the twentieth-century for which the term 'mass' was particularly appropriate' and the sense that the very use of the term 'mass' in an anthropological context, suggesting an unknown which has to be explored, anticipates a broadening of cultural consciousness within the rigid class society of England (Laing 1980: 155–6). Both of these ideas were forcibly argued at the beginning of the Second World War by George Orwell in his twin celebrations of an emergent classless England 'along the naked roads and in the naked democracy of the swimming pools' (Orwell 2000c: 408) and of a new 'proletarian literature': 'So long as the bourgeoisie are the dominant class, literature must be bourgeois. But I don't believe that they will be dominant much longer, or any class either. I believe we're passing into a classless period, and what we call proletarian literature is one of the signs of change' (Orwell 2000b: 297). The question arises as to whether M-O were able to realise their initial goals for social transformation only by working for the State and helping to implement a programme for reform similar to that outlined by Orwell in *The Lion and the Unicorn*, which anticipated many features of the 1945 settlement. Even if that should be the case though, the very fact that the M-O founders, like Orwell, came from the public-school-educated upper middle classes suggests the possibility that all this does is illustrate the ways in which the ruling class were able to adapt to changing times in order to maintain rule.

One of the central theses of this book is that M-O can only be understood in relation to all these major cultural, political and historical themes. While there is undoubtedly truth in the claim that the absence of a comprehensive history of M-O 'makes it hard to develop a chronological and inclusive overview of how they operated' (Sheridan et al 2000: 37), it is also the case that there are sufficient articles providing either short histories or accounts of particular aspects of M-O for the basic details to be readily available (notably Jeffery 1978; Calder 1985; Sheridan et al 2000: 21–42; Highmore 2002a: 75–112). Therefore, it is the significance of M-O that has to be established in the first place so that the need for a comprehensive organisational history – which to be fully comprehensive would probably have to run to several volumes – will become apparent. So, although this work does provide a more comprehensive history than any that has gone before, it should be treated primarily as a combination of cultural history and critical exposition concerned as much with the ongoing value of M-O today as its origins in the 1930s. Of course, this does not preclude the necessity of providing an outline summary of M-O in this introduction so that readers can have some understanding in advance as to why certain frameworks, contexts and aspects are chosen for analysis in the main body of the book and how they fit together in the greater argument. To this end, the following brief chronological account of M-O, the history of which will be covered in much greater detail in the main body of text, is structured to show how their self-understanding changed across time.

A Brief History of Mass-Observation

The existence of M-O was first announced under the heading 'Anthropology at Home' in a letter to the *New Statesman* published on 30 January 1937 and signed by the three founders: Tom Harrisson (1911–76), Humphrey Jennings (1907–50) and Charles Madge (1912–96). Harrisson, a schoolboy ornithologist who had turned into an anthropologist during the course of four international scientific expeditions, was enjoying fame as the author of *Savage Civilisation*, a polemical account of his experiences with 'cannibals' in the New Hebrides (now Vanuatu), published earlier that month. Jennings was a documentary film-maker, painter, set-designer and surrealist, who had sat alongside his friend André Breton on the organising committee for the International Surrealist Exhibition held in London the previous summer. Madge, besides being a reporter for the *Daily Mirror*, was a communist, a poet and a regular contributor to the influential journals *Left Review* and *New Verse*. Their joint letter made reference to this diverse range of disciplinary backgrounds: 'The artist and the scientist, each compelled by historical necessity out of their artificial exclusiveness, are at last joining forces and turning back towards the mass from which they had detached themselves' (Harrisson, Jennings and Madge 1937: 155).

This joining of forces was an amalgamation of two pre-existing projects. In November 1936, having recently both lectured to the Royal Geographical Society and appeared on the newly-launched BBC television service, Harrisson was working eleven hours a day in a cotton mill in Bolton as a form of participant observation. He was later to justify this choice of location by arguing:

> The one and only thing which I could find that affected the lives of people in all the places I had been to everywhere in the world was the Unilever Combine. Even the cannibals in the mountains of Melanesia were touched by the tentacles of this colossus, buying copra, selling soap. Unilever stemmed directly from William Lever. He was born and started business in Bolton. So I followed there (Harrisson 1959: 159).

This one man operation would quickly be transformed under the auspices of M-O into a huge study of every aspect of life in 'Worktown' as it was named in emulation of Robert and Helen Lynd's classic American study, *Middletown*. Meanwhile, Jennings and Madge had spent 1936 toying with ideas for 'Popular Poetry': a surrealist-inspired social movement which would map the collective mass consciousness of the nation through the establishment of factory- and college-based 'Coincidence Clubs'. By December, they had with a group of others produced a questionnaire for public circulation – a precursor of the monthly directed questions which M-O would subsequently send to its 'National Panel' of volunteer observers – inquiring into such matters as personal superstitions and responses to the abdication crisis and the burning down of the Crystal Palace. Madge wrote about this group to the *New Statesman*, making the enigmatic claim that 'only mass observations can create mass science' and calling for people to take part (Madge 1937a: 12). It was by reading this letter that Harrisson became involved in the project, bringing his own ideas for the study of Bolton with him. The obvious differences between these two projects and the subsequent splits in the movement have led commentators – and the principle participants themselves on subsequent occasion – to suggest that the enterprise was fundamentally divided from the start. However, what needs to be taken into account is the way that these divisions were accentuated by the fundamental social division in Britain between modernity and tradition which, as Harrisson, Jennings and Madge all realised, was laid bare by the abdication crisis. As we shall see, opinion in the country was polarised by a modern king, at ease with air transport and radio broadcasts, supported throughout the crisis by an emergent Woolworths-shopping, Penguin-book-buying modern mass society and opposed by Parliament, the Bishops and the nonconformist North. In practice, this meant that the two projects which formed M-O were actually directed at opposing social forces. What is interesting, therefore, is not that M-O eventually split along these predestined lines, but that it was ever able to transcend such fundamental difference. Of course, this social division was not exclusive to the 1930s as witnessed by the journalist Peter Hitchens's comments on Tony Blair's first General Election victory:

'The two Britains which faced each other in April 1997 were utterly alien to one another and unfairly matched. One was old and dying, treasuring values and ideas which stretched back to a misty past. One was new and hardly born, clinging just as fiercely to its own values of classlessness, anti-racism, sexual inclusiveness and license, contempt for the nation state, dislike of deference, scorn for restraint and incomprehension for the web of traditions and prejudices which were revered by the other side' (Hitchens 2000: xxxv–xxxvi).

Therefore, the value of exploring what was possibly a short-lived late 1930s inclusive cultural consciousness registered by M-O is not simply historical but valuable for confronting ongoing cultural and political concerns in Britain and similar problems created by the onset of modernity across the world.

The potential strains of early M-O were held in creative tension by a fluid tripartite structure in which Madge was responsible for national day-surveys and directives, Harrisson responsible for the Worktown study and Jennings responsible for the presentation of results. This particular organisation was enabled by the collective decision to treat 'images' as the social facts of the investigation. The concept of the 'Image' in the 1930s had a particular modernist resonance that entailed something more specific than a mere pictorial impression. Ezra Pound had defined the Image as 'that which presents an intellectual and emotional complex in an instant of time' (Jones 2001: 39) and the aim of the Imagist poets had been to make the image rather than the word into the unit of signification so that their poems generated their own meanings separate from dominant narrative associations. For Pound, the point was to develop a new school of poetry distinct from symbolism, which he considered dependent on exactly such associations: 'The symbolist's *symbols* have a fixed value, like numbers in arithmetic, like 1, 2 and 7. The imagist's images have a variable significance like the signs a, b, and x in algebra ... the author must use his *image* because he sees it or feels it, *not* because he thinks he can use it to back up some creed or some system of ethics or economics ...' (cited in

Jones 2001: 21). The appeal of these concepts to M-O – influenced as we shall see by their interest in Surrealism – lay in the twin ideas of variable significance and escape from externally imposed associations. Accordingly, a major early goal of M-O was to spread the expertise of painters and poets in the use of imagery to the mass-observers, thus liberating their perceptions from externally imposed sense associations and creating a sense of the possibility of change. This can be seen from the expressed intention of training observers: 'We intend to issue series of images, like packs of playing cards, and to suggest various exercises which can be played with them' (Madge and Harrison 1937: 37–8). M-O's image-based approach is clearly central to the montage sections of their first book *May the Twelfth: Mass-Observation Day-Surveys 1937*, edited by Jennings and Madge, but it also features in much of the early Worktown activity directed by Harrisson. Not only did Harrisson instigate projects involving artists such as Julian Trevelyan and William Coldstream, but he also encouraged written reports such as 'The Bolton Tortoise', reprinted at the start of *Speak for Yourself*, which demonstrated an imagist quality: 'Large, tough guy with masses of hair held down by a hairnet ... suddenly takes a small live tortoise out of his overcoat pocket and threatens woman with it' (Calder and Sheridan 1984: 1–2). The initial aim was to provide social facts that could not simply be reduced to statistics.

However, Jennings's departure from M-O in late 1937 and the commercial failure of *May the Twelfth* led to a shift in the balance of the organisation and a change of priorities. In courting anthropological respectability in general and the patronage of Bronislaw Malinowski in particular, M-O switched to treating their observers as informants in the ethnographical sense and redefined their purpose: 'Beyond the sphere of law and legal contract is the sphere of custom and agreement: a sphere of unwritten laws and invisible pressures and forces. The function of Mass-Observation is to get written down the unwritten laws and to make the invisible forces visible' (Madge and Harrison 1938: 8). This approach resulted in the successful Penguin Special *Britain by Mass-Observation*, famous for its account of the Munich crisis but also highly significant for its detailed study of the 'Lambeth Walk' dance craze.

In August 1938, Harrisson and Madge switched their respective locations of Bolton and London (partly because of problems in Madge's personal life) as the original Worktown study of pubs, pools, religion and politics was wound up to be replaced by the Madge-led 'Social Factors in Economics' project, with Harrisson taking over what was now called the National Panel. The economics project, which ran from September 1938 to September 1939, was an investigation of the social psychology of working-class saving and spending, which employed trained social psychologists such as Denis Chapman and Gertrud Wagner and provided the model for much of the governmental social research undertaken during the Second World War by the Wartime Social Survey, in which former M-O researchers were key participants. In 1940, Madge devised and conducted social surveys into saving and spending for John Maynard Keynes in the run up to the 1941 Budget before leaving M-O and going on to work with, first, William Beveridge on industrial planning and, later, Michael Young at PEP. The under-standing Madge gained of the social needs of the working class through his M-O research into patterns of saving and spending illuminates the thinking behind the formation of the Welfare State: 'I don't think that money is the sole driving force of working-class activity ... Prestige and social status are sought, even at the expense of economic needs' (Madge 1941a: 37).

During the Second World War, M-O, under Harrisson, initially worked for the MOI – who were attracted to M-O on the strength of the analysis of public opinion during the Munich Crisis in *Britain* – and later Naval Intelligence, collecting material on public morale, includ-ing such areas as responses to political leaders and behaviour in cinemas. Anticipating the focus groups of the 1990s, Harrisson claimed that M-O could supersede standard opinion polling – then, of course, a fairly recent development – because conventional interviewing did not give information about what people were thinking but only about what they were prepared to say to a stranger. Citing M-O's ability to distinguish between what a person says to a stranger, an acquaintance, a friend, his wife, himself and in his sleep, he concluded: 'It is at the level of wife, self and dream that the most honest assessment of morale can be made. From the private opinion of 1940 comes the public opinion of 1941' (Harrisson 1940c: 1). This claim was arguably vindic-ated by Harrisson's prediction of the postwar Labour victory as early as 1943, as subsequently recorded in the January 1944 issue of *Political Quarterly*: 'Some months ago I remarked, at a research meeting, that social surveys suggested a probable Labour victory, by a wide margin at the next election, if Labour played now for success' (23). As he made clear, this analysis was based on M-O diaries, letters and talk rather than public statements to opinion pollers. By this point of the war, Harrisson had been conscripted and shortly afterwards departed to serve behind enemy lines in Borneo as a Major in the S.O.E.

On his return to Britain and M-O (run in his absence by Bob Willcock) in 1946, Harrisson wrote a 'Demob. Diary' for the *New Statesman* complaining about how Britain was already characterised by a 'pathetic nostalgia' for 'masculine adventure' as everyday life sank back into 'the restricted circle of private western experience' (221). A further shock on his return had been to 'find sociology marking time on 1937'. This observation was made in a provocative article published in *Pilot Papers*, a seminal Cultural Studies journal edited by Madge. Distressed by the 1946 governmental Clapham Report on the provision for social and economic research, which advocated that social research be carried out exclusively by universities and governmental departments, Harrisson attacked what he called a 'statistical obsession' and prophetically warned that without qualitative work 'sociologists will never be able to look further than the day after tomorrow' (Harrisson 1947a: 16). Furthermore, he accused those academic sociologists insisting on statistically accurate and representative research of merely rationalising their own inability to integrate with life in a mass society. It is difficult to see this as anything other than a swansong for both M-O and the modern mass society that appeared to be developing in the late 1930s. Although both Harrisson and Madge were involved in trying to establish an alternative social science forum at PEP, the vision of British sociology laid out in the Clapham Report was to prevail and presented a stark choice to researchers. Unable to accustom himself to the dullness of postwar Britain, Harrisson had already accepted the colonial posting of Curator of the Sarawak Museum which he was to hold for nearly twenty years. Madge took the other pathway and in 1950 became Professor of Sociology at Birmingham University. After 1949, M-O, with no remaining direct input from its founders – although Harrisson retained financial interests – became a limited company devoted to commercial market research.

The sheer weight of qualitative material collected by M-O in this initial period between 1937 and 1949 remains unparalleled. Aside from the day-surveys, completed by volunteers on the twelfth day of each month throughout 1937, there are the results of three years of intensive participant observation in Bolton including literally millions of words on religious, political and leisure institutions. There is an extensive national collection of material on popular culture, advertising, parliamentary byelections and particularly on the public response to the onset of war. Diaries kept by panel members during the war – some of which such as *Nella Last's War* and Naomi Mitchison's *Among You Taking Notes* have since been successfully published in their own right – collectively amount to a comprehensive documentation of everyday social life. There is also the extensive work carried out for the MOI and the detailed studies of industrial production which underpinned M-O books such as *People in Production* and *War Factory*. There is a further mass of material from after the war including detailed studies of demobilisation and the birth rate.

It is this sheer informational value of the project that enabled it to survive. The huge collection of papers would have mouldered away in the cellars of Mass-Observation Ltd if not brought to attention by the research of historians Angus Calder and Paul Addison. This led eventually to Asa Briggs, the Vice-Chancellor at the University of Sussex, providing a home for the collection and inviting Harrisson to take up an academic position in order to sort it out. The resulting public archive was opened in 1975 (Sheridan et al 2000: 38) shortly before Harrisson's untimely death in a road accident. However, the positive reception enjoyed by his posthumously published collection of wartime M-O reports, *Living Through the Blitz*, combined with the success of Calder's *The People's War* and Addison's *The Road to 1945*, legitimised the archive as one of the key sources for social historians of wartime Britain and helped ensure its continued existence. The continuing presence of what was now called the Tom Harrisson Mass-Observation Archive eventually prompted the social anthropologist David Pocock to restart the National Panel in 1981 in order to monitor reactions to the Royal Wedding of that year and record social conditions under the Thatcher Government. Directives have been continuously sent out and replied to ever since, with the resultant material available for research purposes. The current incarnation of the Archive and Project is in very good standing under the direction of Dorothy Sheridan, who has worked there since before Harrisson's death.

Everyday Life and Social Transformation

However, there are other ways of viewing the M-O A than simply as a primary source for researchers. At around the same time as Harrisson was venting his frustration with newly postwar Britain, the great French social theorist Henri Lefebvre was confronting the rapid decline of post-liberation optimism into the sterile attitudes which would eventually come to dominate during the Cold War period. His response was a *Critique of Everyday Life*, published in 1947, in which he called for 'the undertaking of a vast survey, to be called: *How we live*' including examination of 'the details of everyday life as minutely as possible – for example, a day in the life of an individual, any day, no matter how trivial' (196). As we know, this search for 'unconscious social mechanisms' had been anticipated ten years before by M-O. Therefore, there are strong reasons for considering M-O within the growing sub-discipline of 'Everyday Life' Studies – a position that has been taken recently by Ben Highmore's *Everyday Life and Cultural Theory*, in which M-O is awarded equal consideration to such theorists and movements as Lefebvre, Georg Simmel, Walter Benjamin, Michel de Certeau and the Surrealists. Highmore's claim that such 'avant-garde sociology ... is fashioned when the everyday is taken as the central problematic' (22) provides a theoretical and historical context for reconsidering earlier attempts to represent M-O as sociological innovators, most notably Nick Stanley's unpublished – but known to a generation of M-O A users – Ph.D thesis, which made a brave attempt to argue that M-O's literary and aesthetic techniques lent an 'extra dimension' to the practice of the social sciences (N. Stanley 1981; see also Chaney and Pickering 1986: 29–44). The successful establishment of an everyday life approach would have huge significance for the reception of M-O, because hitherto their innovative interdisciplinary practices always fell between academic schools so that literary critics have been able to claim that their books failed because 'it is the *mass* in Mass-Observation which is numbing' (Hynes 1982: 286) while sociologists have been able to denounce that 'as social research, the methods used were unsystematic, relied too greatly on externals, and suffered from the investigators' conception of what they were doing as a form of art' (Bulmer 1985: 11).

Much more importantly, though, looking at M-O from the perspective of Everyday Life studies allows us both to see its wider cultural and historical significance, particularly in relation to its contemporary European movements, and to understand theoretically how it could pursue aims of social therapy and transformation. For example, it is possible to see Lefebvre's working concept of everyday life, in which the repeating units of capitalist exchange – money, time etc – comprise an everydayness that renders everyday life into empty time that has to be filled, while historical memory is simultaneously reduced to a trace existence, as owing clear debts to Freud's notion of spatial-temporal consciousness being produced by banishing memories to a trace form (see Freud 1984b: 295–301). Potentially, therefore, a therapeutic politics of everyday life would not be dissimilar to Freud's model of the analytical encounter, in which the trace is brought into contention with consciousness in a dialectical process of remembering, repeating and working-through which causes the patient to 'act' out what has been repressed as a 'piece of real life' (Freud 1958a: 150–2) and, therefore, find a safe outlet for otherwise disruptive emotional impulses. This analytic process contains an inherent ambiguity for in order for it to function it has to override the pleasure principle, which is the primary defence mechanism for protecting the ego from trauma by regulating the level of excitement and agitation resulting from external or internal stimuli to as low and stable a level as possible. The pleasure principle carries out this function by discharging pleasurable stimuli – as exemplified by the sexual act which is in effect only the 'momentary extinction of a highly intensified excitation' (Freud 1984b: 336–7) – and repressing unpleasurable stimuli. Therefore, the patient can only be made to act out the traumatic events in the analytic encounter by circumventing the pleasure principle's repression. This circumvention is enabled by inducing a regression from the pleasure principle to the infantile stage of ego defence mechanism which developmentally precedes it: the compulsion to repeat. Here, the classic example is the child who compensates for the mother's absence by staging the disappearance and reappearance of toys within reach. This repetition of a traumatic event allows the child to attain an active role in place of the usual passive one and the stage as a whole helps to form the pleasure principle because the lines of cathexis (neural links) which allow emotional discharge can only be formed by repetition. However, while this demonstrates that therapy is successful to the extent it is able to generate a new line of cathexis by inducing a controlled compulsion to repeat, the point of ambiguity arises from the question of why the passive maintenance of the ego by the pleasure principle should be seen as preferable to the active processes unleashed in the analytical encounter? Consciousness is momentarily set free to act out 'reality' before being returned to a state of conformity with external reality – the reality principle – but why is the one reality privileged over the other? In fact, the Freudian analytical encounter offers a model for transformation to creative and critical thinkers – hence its continuing interest despite it no longer being at the forefront of modern psychology – in which the 'reality' a person is aligned to can be transformed, either externally and within proscribed limits as a form of therapy or internally and without limits in what would amount to the development of an unfettered full consciousness.

Great public crises such as those surrounding the abdication and the 1938 Munich agreement between Chamberlain and Hitler, could be seen as analogous to individual trauma and as arising because the normal societal defence mechanisms could not cope with the shock of the disturbance. M-O's interventions in these crises should be seen as attempts to induce social transformation, although whether this went beyond social therapy in either aims or achievement is a much more difficult question to answer. In later life Madge acknowledged the importance of Freud's *Psychopathology of Everyday Life* to M-O and its influence on their early concept of the 'coincidence'. Freud's book was widely available in English in the 1930s, even appearing as a mass-market paperback edition in 1938 under Penguin's Pelican imprint. Freud's argument was that at the 'back of every error [whether slip of the tongue, memory loss or accident] is a repression' (161) of unconscious will or desire so that the errors happen precisely because 'unconscious thoughts find expression as modifications of other thoughts in unusual ways and through outer associations' (213). Clearly, there are similarities between the mechanism by which the Freudian unconscious disrupts dominant narrative associations and the conscious practice of the Imagist poet as previously discussed. Therefore, as we shall see, there is a strong case for arguing that Madge and Jennings, following the influence of the Surrealists, chose to read Freud against the grain and so set out by collecting images from their trained volunteer observers to consciously create the conditions for unfamiliar associations that would allow the (possibly collective) unconscious greater opportunity for expression than would otherwise 'naturally' occur and so accelerate social transformation. In assessing exactly which changes in society can be attributed to M-O, whether independently or in conjunction with other social forces, there are a number of good reasons for agreeing with Alan Read that '... the formation of an idea of nationality, the everyday practices which make up "Britain" itself, owes something very distinctive to the work of the "Mass-Observation" movement' (A. Read 1993: 70).

Firstly, in *May the Twelfth*, M-O anticipated Habermas's arguments concerning the structural transformation of the public sphere, by showing how the advent of modern mass society had empowered the media as the prime determinant of all social relationships. Their demonstration of the dilution of the historical relationship between the monarch and the people into just one – albeit, as they phrased it, the archetype – among a chain of media relationships foretold an essential component of the postwar state.

Subsequently, *Britain* utilized the ideas on pastoral of their friend, William Empson, to portray the Lambeth Walk as providing a model for an alternative myth, that could be deployed against the threat of

Fascism, in its capacity to represent Britain as a heterogeneous but nonetheless unified nation. In this manner, M-O introduced the pantomimes of class which were to characterise British society for the next fifty years.

Thirdly, as already mentioned, the M-O founders went on to gain central influence in the wartime British State – Harrisson with the MOI, Jennings with the Crown Documentary Unit and Madge with Keynes, Beveridge and PEP. In particular, Madge devised and conducted the social research which won Keynes and the Treasury over to the income tax option for funding the war deficit. As a direct result of what began as M-O research, the 1941 Budget extended income tax 'downwards' to four million working-class payers for the first time, fundamentally altering the structural relations of prewar British society and laying the necessary foundations for the postwar Welfare State. The presiding Chancellor of the Exchequer, Sir Kingsley Wood, made the famous claim that 'the Englishman has a genius for co-operating with the tax collector' as a triumphant conclusion to an M-O style montage of direct quotations – incorporated in his budget speech to allay fears in the House of Commons over working-class willingness to pay – compiled by Madge (Wood 1941: 1301–2). This unprecedented registering of working-class voices at the core of the State can be seen as the culmination of M-O's audacious political project to replace the space of former England with a British mass democracy.

Yet it is also possible to view the transformative project of M-O as a strictly limited exercise in social therapy: one which might even be pejoratively labelled social engineering. In 'Magic and Materialism', a 1937 *Left Review* essay in which Madge explains M-O in Marxist terms, he seems to be advocating a mass therapy which will lead to the collective abandonment of the pleasure principle in favour of the acceptance of the evolutionary maturity of the reality principle (in which immediate pleasure is deferred in the interest of long-term gains) – a reality that he links with Socialist Realism in art and literature as much as science. According to *Britain*, pleasure in the shape of popular culture (with the singular exception of the Lambeth Walk) is the practical shape in which Fascism invades the politically isolated home life of the masses, as the racing news and daily horoscopes distract readers from the news about Czechoslovakia. M-O's *War Begins At Home* is nakedly authoritarian: 'Effective government and leadership has as a prime obligation, and also as a necessity for its own survival, the job of telling every citizen who isn't in a lunatic asylum roughly what to do and roughly what to think about the issues which affect everybody' (Harrisson and Madge

1940: 49–50). It is clear from Madge's writings at the time of the 1941 Budget that he regarded the obvious desire of many of the Northern working class to leave their home towns and move south to improve their standard of living as a form of collective neurosis. One of the results of that Budget, which effectively redistributed wealth within the working class, was to obviate the necessity for such an exodus and slow down social mobility in general. On the basis of this evidence, it is possible to argue that the social psychology of M-O was close to the governing social psychology of the eventual postwar Welfare State: the adaptive ego system in which society is scientifically ordered to help people towards the reality principle by planning for full employment and providing benefits to keep those temporarily unemployed on the straight and narrow.

Rather than seeing M-O as fluctuating between these two approaches, it probably makes more sense to see these apparent contradictions as the result of an attempt to maintain a reasonably stable society while simultaneously working for a planned transformation of that society. Such aims can be readily detected in Madge's postwar work in sociology, notably in his 1964 book *Society in the Mind*. Here, Madge described the social eidos – the mental framework of society – of the Sixties as predominately rational-technical and suggested the need of relaxing these socially coercive bonds in favour of individual aesthetic development. However, this advocacy of a combined social therapy and transformation within the parameters of the postwar state was swept away by the events of 1968, which coincided with Madge's tenure as the Dean of Faculty of Social Sciences at Birmingham. Distressed, Madge left his academic post and ended up reaching a complete state of physical and mental crisis by the mid 1970s, as well as total disillusion with sociology, Marxism and any form of Leftism.

The dominant academic conception of the new project – as expressed in *Writing Ourselves* by Sheridan, Brian Street and David Bloome – is of M-O as a practice of 'Writing Ourselves and Writing Britain'. Here, life writing is seen as a Foucauldian reverse discourse in which people use the materials provided by dominant institutions against those dominant institutions, to cut out a certain space to live in. Even an act as straightforward as making a list on the back of an envelope is a way of cutting a certain space for human agency out of the relentless passage of calendar and clock time. As M-O recognised in 1937, such apparently mundane activities as list making and diary writing are forms of popular poetry, as committed as the most intractable modernist verse to resisting the dominant meanings imposed by society. Promoting the use of images was intended to enable that popular poetry to move beyond mere resistance to the possibility of signifying new meanings. Not just the new significations of an artistically and intellectually gifted minority, but new significations of the masses by the masses for the masses. This is not to suggest that the trajectory of M-O over the last sixty-eight years has been circular, but that the founding spirit is still relevant to the present-day project.

Inversely, present-day perspectives on life writing provide a significant filter for considering the past activities of M-O. It is interesting to note in retrospect how the writing of some observers can be read as not just resistant to dominant ideologies but to the leaders of M-O themselves. The following comments on *War Begins At Home* show how some observers were clearly not concerned about the book's ostensible purpose of demanding active leadership of the bewildered masses: 'Before we caught the bus home we went in the town library to ask if they had got 'our' book. (We always call it 'ours', hope M-O doesn't mind, but you see we've never had anything we've written in print before and claiming 14 lines and J. [the diarist's sister] 25 lines we feel a proprietary interest in the publication, and that everybody ought to sell and read it)' (cited in Sheridan 2000: 88). The presence of these resistances indicates how the idea of M-O – the promise of the emergence of a modern mass society – was greater than the positions taken by its founders. Yet in order to highlight these resistances it is necessary to focus on the characters, careers and ideas of those founders, who so neatly personalise the social forces and tensions of not just the movement, itself, but also the wider dynamics of the long twentieth century. After all, while it is surely right to reject the notion that history consists of the lives of great men, it is equally wrong to forget that history is driven by human agency, even if the circumstances are not freely chosen.

1
Historical Background

Everyday Life in the Long Twentieth Century

The question of how we periodise the twentieth century is not an idle one but crucially important to both the original and the current incarnation of M-O and the wider field of Everyday Life studies. According to some historians, the period between 1914–1989 was an historical anomaly – at its most extreme, such a view entails seeing this 'short twentieth century' as conterminous with a world civil war between Bolshevism and the Liberal West with the latter emerging triumphant victors – so that the fall of the Soviet Union and the consequent global political realignments have allowed events to get back on the course they were following before the First World War. Such views do not by any means preclude historical studies set within that period, but they do preclude studies – such as this one – which seek to demonstrate the ongoing valency of ideas and movements of that period within the present. The decade of the 1930s in which M-O originated is no longer celebrated as an extraordinary fertile moment in which the world lay on the cusp of social transformation but as 'the dark valley of Depression' (Brendon 2001: xiv). In consequence, it seems radical politics have become restricted to such a narrow cultural sphere that everyday practices such as reading and cooking can be celebrated as major acts of resistance and recuperation. As John Roberts has argued, contemporary theories of cultural studies appear to confuse 'the social function of "storytelling" from below ... for the work of emancipation itself' (J. Roberts 1999: 27). In order to challenge this state of affairs, it is necessary to establish a longer continuity, cutting across the so-called anomalous period at the heart of the twentieth century. While the concept of the long twentieth century originates in economic history, the following discussion extends beyond this discipline to establish a broad framework of social, cultural and political trends.

Giovanni Arrighi's *The Long Twentieth Century* develops Fernand Braudel's idea that finance capital is not the high point of global capitalism but a repeated stage marking the transition from one cycle of capitalist accumulation to another. Suggesting that this idea corresponds to Karl Marx's general formula of capital MCM', which represents the transition of fluid money capital (M) into fixed commodity capital (C) and back into (increased) fluid capital (M'), Arrighi identifies four overlapping systemic cycles of capitalist accumulation: respectively long fifteenth-sixteenth, seventeenth, nineteenth and twentieth centuries (6, 364). Periods of finance capitalism, when capital is released from the fixed commodity form, constitute the overlapping transitional phases between the cycles. For example, the financial expansion within the period 1870–1914 can be seen as both the final phase of a long nineteenth century and the opening phase of a long twentieth century because it effectively began with capital flowing out from the fixed commodity form of nineteenth-century industry – typically associated with British global hegemony – and ended with capital flowing into twentieth-century mass production lines – typically associated with American global hegemony. Some economic historians have qualified Arrighi's theory by arguing that economic hegemony briefly passed to Germany as a consequence of scientific and technical advancement in the 1920s before becoming properly consolidated in the United States following the conclusion of the Second World War (see Broder 1999). Not only does this qualification help to explain the influence exerted by German cultural politics between the wars but also, as will become clear, it is essential for understanding everyday life as an historical process. Furthermore, it suggests that the dominant commodity form of capitalism in the twentieth-century cycle cannot simply be characterised as a 'Fordist-Keynesian regime of accumulation' – in which Keynesian economic policies ensured an optimal environment for Fordist production processes leading to regularly increased productivity and mass consumption (Arrighi 1994, 2) – however much that description might fit the postwar period. In any case, since the early 1970s the stability of this Fordist-Keynesian system has broken down and capital has once again flowed out of the fixed commodity form in a new phase of financial expansion.

If the wider process of overlapping systemic cycles of capitalist accumulation were to continue, we might expect this latest phase of financial expansion to mark both the ending of the long twentieth century and the opening of a long twenty-first century. It is clear, though, that a specific periodisation can only be made convincingly in retrospect. For example, it is only possible to talk of a long nineteenth century because we know that different cycles of capitalist accumulation preceded and succeeded it. Until a new fixed commodity form arises, or some other economic change of comparative magnitude occurs, it will be impossible to confidently date the end of the long-twentieth century. This uncertainty leaves the whole concept open to question and is compounded by the presence of a competing theory: the aforementioned concept of the short twentieth century, which is most likely to be familiar to readers from Eric Hobsbawm's *Age of Extremes: The Short Twentieth Century 1914–1991*.

Of course, on the one hand, the historical events are the same however they are categorised but, on the other hand, our understanding of these events is conditioned by the conceptual paradigms that govern our thoughts. It is easiest to show what is at stake in the choice between these two paradigms by outlining some of the consequences of adopting Hobsbawm's position. The kernel of his argument is that: 'The world that went to pieces at the end of the 1980s was the world shaped by the impact of the Russian Revolution of 1917' (4). Here, Hobsbawm cleverly subverts accounts of the 'world civil war', which share the same time frame, by focusing on the great irony that 'the most lasting results of the October revolution, whose object was the global overthrow of capitalism, was to save its antagonist' both by defeating Nazi Germany and by 'establishing the popularity of economic planning, furnishing [capitalism] with some of the procedures for its reform' (7–8). Furthermore, he concedes the general charge of historical anomaly in order to specify that it was the unprecedented period of economic growth and social transformation in the 'Golden Age' of 1947–73 that was truly anomalous. The trouble with this is that while it challenges right-wing ideas of history, it is equally conservative in its own way: seeking rather to close the twentieth century off as an idealised past than to continue further the struggles which characterised it. Such closure positively invites the ideological assertions of 'The End of History', which Hobsbawm's text is ostensibly opposed to, and leaves no defences against a history like Mark Mazower's *Dark Continent*, which claims that the collapse of the Soviet Empire has made possible 'a remarkable combination of individual liberty, social solidarity and peace' (410) in which 'Europeans accept democracy, because they no longer believe in politics' (404).

The almost satirical absurdity of this statement is redeemed only by Mazower's tendencies to undermine his own account, such as in his admission that this depoliticised 'democracy' first emerged in the 1940s – that is, during Hobsbawm's Golden Age – as a consequence of postwar demobilisation: 'The great tide of mass mobilisation began to ebb, and with it the militarism and collectivism of the inter-war years ... People rediscovered democracy's great virtues – the space it left for privacy, the individual and the family' (xi). This suggests that there is nothing new about apolitical democracy. Moreover, this version of democracy can be seen to collapse further back through history, when Hobsbawm's own tendencies towards self-contradiction are highlighted. He refutes his own contention that the wartime victory of post-enlightenment rationality – in the form of a coalition of Communists, Conservatives, Liberals and Social Democrats – over Fascist reaction in 1945 was the start of the Golden Age, by arguing that the Spanish Civil War anticipated both the shape of the forces that were to destroy Fascism and the democracy that was to succeed it:

> Both the Spanish government and, more to the point, the communists who were increasingly influential in its affairs, insisted that social revolution was not their object, and, indeed visibly did what they could to control and reverse it, to the horror of revolutionary enthusiasts. Revolution, both insisted, was not the issue: the defence of democracy was.
>
> The interesting point is that this was not mere opportunism or, as the purists on the ultra left thought, treason to the revolution ...(162–3)

This self-contradiction is reinforced because Hobsbawm has only a few pages before unambiguously denied any significance to the conflict: 'In fact, and contrary to the beliefs of this author's generation, the Spanish Civil War was not the first phase of the Second World War, and the victory of General Franco who, as we have seen, cannot even be described as a fascist, had no significant consequences' (156–7). The net result of this ideological manoeuvring is a rejection of the values of the 1930s generation, tellingly accomplished by the use of the Stalinist term 'ultra left', in order to describe what Hobsbawm admits was 'an essentially defensive tactic' as a 'democracy of a new type'.

Hobsbawm's account of the genesis of modern democracy is strongly reminiscent of Lefebvre's understanding of the origins of modern everyday life, albeit that they display opposing attitudes to the process they describe. As Michel Trebitsch has written: 'For Henri Lefebvre modernity and the everyday are historical categories, and if they cannot be dated precisely, at least they can be located at a moment of fundamental historical trauma: the failure of revolution, which was completed, at the very moment of the world crisis, by the advent of Stalinism and Fascism' (Trebitsch 1991: xxvii). Therefore, it is not surprising to find that the condition of 'new democracy', as variously described and periodised by Hobsbawm and Mazower, was also highlighted in contemporary accounts of Nazi Germany: 'The masses which were so nationalistic before, under the strain of everyday life under the Nazis and of a surfeit of lying propaganda, have become largely indifferent to all the bigger political issues, and completely immersed in their petty everyday affairs' (Borkenau 1940: 43). It is a greater shock to see similar trends identified in the Soviet Union even before the full onset of Stalinism, as in Walter Benjamin's prescient analysis of 1926: '... domestically ... [the Soviet government] is above all trying to bring about a suspension of militant communism, to usher in a period free of class conflict, to de-politicise the life of its citizens as much as possible' (Benjamin 1986: 53). Moreover, earlier accounts of a depoliticised everyday life are to be found in works by Georg Lukács and Martin Heidegger that predate the Russian Revolution, which suggests that the tendencies described by Lefebvre already existed before the point of historical origin that he identified and, in turn, that the key events of the 1930s were always predictable, even if not the only possible outcomes.

The idea of a history of everyday life conterminous with the long twentieth century is supported by Jürgen Habermas's highly influential work, *The Structural Transformation of the Public Sphere*, which describes the onset of 'neomercantilism' (state intervention) and 'refeudalisation' (reversal of the separation between state and civil society) as marking the depoliticisation and 'downfall' of the public sphere 'from the time of the great depression that began in 1873' (141–3). According to Habermas, the bourgeois public sphere had first come into being with the amalgamation between 'court' and 'town', which brought together a public display of authority descended from feudal society with a civil society separated from the state because based on the economics of private property. Thus, the term public no longer simply designated all those who were subject to public authority as in the early Middle Ages but came to indicate an increasingly autonomous sphere. The development of print culture facilitated the exchange of ideas and enabled the existence and enlargement of this public sphere composed of 'private people engaged in rational-critical debate' (106–7, 117). In particular, the widespread expression and discussion of private opinions was collectively manifested into the new historical phenomenon of public opinion. Famously, Karl Marx denounced this public opinion as false consciousness because it masked the unified class interests of the property owning bourgeoisie. Nevertheless, the continued accession during the nineteenth century of those without private property to public debates, allowed Marx to argue that civil society would ultimately dissolve as the basis of autonomy switched from private property to the fulfilment of the functions of a self-justifying public sphere in which 'private persons came to be the private persons of a public rather than a public of private persons' (128). However, as Habermas argues, Marx's early vision was never fulfilled because of its reliance on an idea of 'natural order' that did not account for the fact that the extension of political rights proved to be compatible with the continued existence of class society. This apparent paradox was partly assisted by the development of representative democracy, which abandoned the idea of situating political legitimacy in the existence of a 'critically debating public', in order to establish a mediating authority comprised of 'materially independent citizens' (136). The justification for this development, as put forward by John Stuart Mill, was that it released the individual from what was now seen as the tyranny of public opinion or, more particularly, the tyranny of the opinion currently dominant in a public sphere split between unreconciled class interests. As a consequence, the public sphere began to break down because even as it expanded with industrialisation in the second half of the nineteenth century, its political function declined from that of a critically debating public into that of a mass electorate. Increasingly, public authority came to be inextricably linked with emergent state institutions and legislation (health, education, employment etc.), creating a universal social sphere in which the formerly distinct concepts of 'public' and 'private' were merged.

The above process was complemented by an ongoing transformation of work into types of social labour which corresponded to this blurring of public and private spheres. A new form of social experience came into being that was shared between those who would formerly have been described as workers and those who could formerly be said to have exercised control over the means of production. That is to say that a management process, largely separate from the exercise of property rights, emerged as commercial success became increasingly depen-

dent on the internal functioning of the social sphere and correspond-
ingly independent of the capital market. At the same time, the category of employees came into being to describe those who could no
longer be defined simply by the work they did, or the fact that they
sold their labour, so much as by their 'functional performance' as part
of a social organisation. Understanding this as a double movement
helps make clear how the public and the private merged: from the
perspective of the property owners this occupational sphere represented a deprivatisation because it eroded the economic primacy of
their private property; from the perspective of employees the occupational sphere represented a privatisation because it removed them from
a position of being totally subject to their capitalist employers and
instead instituted 'a psychological arrangement promoting the human
relations on the job that create a pseudo-private well-being' (154).
The family was similarly recast by the merging of the public and
private spheres as a process of deprivatisation resulting from the spread
of social legislation was offset by a new found private autonomy con-
ferred by the function of consumption: the two processes combining
to create an entity defined by the consumption of a variety of public
services. Habermas concludes: 'As a result there arose the illusion
of an intensified privacy in an interior domain whose scope had
shrunk to comprise the conjugal family only insofar as it constituted a
community of consumers' (156). Looking back, we can now see how
Mazower's definition of the great virtues of democracy as 'the space it
left for privacy, the individual and the family' is the product of a misconception: that privacy is only a pseudo-privacy and not separable in
any meaningful way from a social sphere in which private and public
functions are merged. More importantly, it can be argued that the
depoliticised condition of everyday life we have seen described
variously by Mazower, Hobsbawm, Lefebvre, Borkenau, Benjamin and
Lukács is a consequence of the structural transformation of the public
sphere and the underlying condition of the long twentieth century.

Although this structural transformation affected the Western world
in general, Habermas's account highlights the extent to which its
forces were particularly concentrated in the newly unified Germany,
because there was not the same well-developed parliamentary constitutional state as in France, Britain or the USA. Without traditional
checks, it was easier for German state functions to dominate what was
in any case a relatively restricted public sphere and to create the social
conditions for rapid industrial development, which soon began to out-
strip its competitors. This process of developing the social sphere was

exemplified by the actions of the German Chancellor, Otto von Bismarck, whose Socialist Laws of 1878 suppressed the free association of German socialist and labour organisations for the next twelve years, even as he introduced Social Security Insurance to create what is often described as the original Welfare State. As a consequence of this accelerated socialisation, acute and distinctive forms of sociological thought developed in Wilhelmine Germany. The key concept of *Verstehen*, the interpretative understanding of social life as developed variously by Wilhelm Dilthey, Heinrich Rickert, Georg Simmel and Max Weber (see Outhwaite 1986), emphasised understanding society from within by means of a developed version of the everyday understanding that all individuals need to function in the social sphere.

In particular, Simmel's description of the ambiguous situation of the individual in the modern world can be seen as the point of origin for modern theories of everyday life, as Highmore has convincingly argued: 'For Simmel, individuality is the dominant mode of experience in a money economy and it is essentially ambivalent. It is lived as atomized, alienated, uniqueness at the same time as it is experienced as uniformity and monotony' (Highmore 2002a: 41). In Simmel's essay, 'The Metropolis and Mental Life', this ambiguous situation is described as the consequence of the capitalist everydayness of modern urban society in which individualism is both the necessary quantitative condition, because classical liberal ideology established the interchangeability of humans as individual units within social and production functions, and the necessary qualitative consequence, because it was only in commodified areas such as taste, style and fashion that a personal aspect to existence could be registered. However, Simmel does not denigrate this modern situation but identifies it as one of the 'great historical structures in which conflicting life-embracing currents find themselves with equal legitimacy'. Such was the importance of this crossroads in 'the world history of the spirit' that the primary intellectual imperative was the need to understand exactly what was at stake: 'In the conflict and shifting interpretations of these two ways of defining the position of the individual within the totality is to be found the external as well as the internal history of our time' (Simmel 1971: 338–9). To aid this necessary understanding, Simmel developed a sociological aesthetics which focused on the link between lived experience and the cultural expressions of that experience. As Highmore observes: 'In this regard, impressionism ... was both the name for the neurasthenic experience of everyday life and the cultural form most suited to representing it' (37). However, the analysis of cultural representation was not simply a means to understanding society but potentially a vehicle for social transformation as acknowledged by Simmel's analysis of socialism in aesthetic terms (analogous to Habermas's analysis of socialism in terms of the public sphere, which we have already considered): 'that society as a whole should become a work of art in which every single element attains its meaning by virtue of its contribution to the whole' (cited in Highmore 2002a: 40).

Simmel's ideas of everyday life as a site of transformation were subsequently strengthened in Germany by the success of the Russian Revolution, which caused Lukács, for instance, to revise his earlier derogatory account of everyday life. However, these revisions also crucially altered the understanding of how this transformation would take place. In Lukács's *History and Class Consciousness*, everyday life is implicitly identified in terms of the theory of reification, so that everydayness is accounted for as the naturalisation (as second nature or false consciousness) of the historically contingent social relations of capitalism and can thus be contested by the everyday resistance of a class-conscious proletariat. While this still renders the transformation of everyday life an immanent rather than external process, it does diminish the fundamental ambivalence that Simmel located across the structures of everyday life by tending to link revolution to the everyday and reaction to everydayness. As the 1920s passed and the likelihood of proletarian revolution in Germany appeared increasingly less likely, there developed a viewpoint seeing depoliticised everyday life as the default condition of a modernity only occasionally punctuated by short bursts of revolutionary resistance. Therefore, cultural politics rapidly assumed importance as being a means of triggering transformation that could not be achieved by social action alone. However, as a consequence of the separation of the elements of everyday life, there now existed a split approach to such politics divided between those such as Walter Benjamin and Siegfried Kracauer who followed Simmel's sociological aesthetic stance of analysing everyday life in its totality, and the dominant position of *die neue Sachlichkeit* which focused on heightening the representation of proletarian experience in order to contest capitalist everydayness, but could also be said to have fallen into the trap of accepting the categories of everyday life as reality and trying to base a politics of praxis upon them.

Neither were these cultural positions any longer simply restricted to Germany: the early onset of structural socialisation and continued successful economic development ensured that, despite the setback of the First World War, Germany was at the forefront of technical and cultural advances and a hegemonic exporter of such ideas. While this unstable situation was ultimately foreclosed by the Nazis assuming power in 1933, it did culminate in one last great cultural dispersal as the leftist and Jewish intelligentsia were forced into exile from 1933 onwards. This is, in part, what makes the 1930s such a nodal point of the long twentieth century because high levels of international cultural crossover were combined with a major shift in global economic hegemony in a dynamic process which was only stabilised after the end of the Second World War. These overall transitions in global hegemony across the long twentieth century are reflected by Habermas's *The Structural Transformation of the Public Sphere*, where he supports his discussion of the period before 1870 with English examples, then uses German examples to illustrate the core of his argument but refers to the USA when looking forward to postwar developments.

One consequence of employing Habermas's analysis in this manner, however, is that it does necessitate a rethinking of the economic basis of Arrighi's model of the long twentieth century. As Habermas makes clear, the social contracts of the structurally transformed public sphere would have appeared as pseudo-contracts (151) according to the classical model of political economy because they were no longer freely entered into but represented a division of necessary social functions. He concludes that the classical model had not described the rules of economics as such but only the functioning of the system under particular historical circumstances (144). This was also the contemporary response of a number of economists to the conditions of the late nineteenth and early twentieth centuries. The development of what was to become termed the neoclassical approach switched emphasis from production to consumption. In particular, the exchange value of goods was no longer seen to be dependent on labour time as in Marx but on marginal utility. In this formulation, exchange values are purely dependent on the relative subjective valuations of the goods by the parties to the transaction with the consequence that, in theory, exchange should be possible that fully satisfies all parties. However, the exchange values will fluctuate as the subjective valuation fluctuates, so that when an immediate need for a good has been satisfied, the first party will no longer place a high subjective valuation on it and the exchange value will decrease to the point at which further exchange will be of no benefit to either. In commercial terms, market prices are achieved when all available goods can be sold without queues of people in front of the shop. The principle is summarised in the form of the Pareto or welfare optimum: 'the optimum of exchange will be achieved when the marginal utility of two goods is the same for all people who want these goods.' This optimum can be seen as one of the rules governing the functioning of the newly developing social sphere in the late nineteenth century.

The development of such neoclassical economic systems has two related consequences for the model of the long twentieth century. Firstly, the emphasis on subjective valuation helped elevate individual and social psychology above economics as a means of explaining the functioning of society. As Lawrence Birken points out, 'If the marginalists were ready to concede that the entire economic system as they saw it was but a means to satisfy the desires of individual consumers, sexual scientists such as Kraft-Ebing and Freud were intent upon finding the laws that governed those desires' (cited in Trotter 1993: 14). Therefore, twentieth century politics cannot be equated with the expression of economic interests in the manner that Marx (or, at least, a particular reading of Marx) was able to characterise the nineteenth century.

Secondly, the designation of the Fordist-Keynesian regime of accumulation as the dominant commodity form of capitalism in the long twentieth century is called into question because such a system is not the most logically optimal form of international trade, which is one reason why it has been subsequently supplanted from its hegemonic position by globalism. Indeed, Hobsbawm is clearly right from this perspective in designating the postwar Fordist-Keynesian period as anomalous. However, the very existence of this anomaly suggests that social psychology was influencing economics. The existence of such a relationship is supported by the social psychological framework applied by M-O's Charles Madge to the research he carried out for Keynes in the run up to the 1941 Budget. Of course, M-O with its self-declared aim 'to get written down the unwritten laws and make the invisible forces visible' (Madge and Harrisson 1938: 8) was itself the very exemplar of that direct consequence of the central determining role of social function and psychology in the long twentieth century: the rapid development of social surveys.

Social Surveys: From Booth's *Life and Labour of the London Poor* to Kracauer's *Die Angestellten* and the Lynds' *Middletown*

The economic success enjoyed by Germany following its structural transformation at the beginning of the long twentieth century derived partly from the fact that its social legislation and welfare provisions maintained and enlarged a domestic market for its industrial and consumer products. In Britain, where the ideas of classical Liberalism had held sway across a strongly-developed public sphere and there was no equivalent political position to the one occupied by Bismarck, structural transformation was inevitably a more reactive and piecemeal process. Most key social legislation, such as the introduction of old age pensions, health and unemployment insurance, labour exchanges, free school meals and minimum wages in sweated industries, was introduced only after 1900 following extensive social research. While there was a long tradition of individual exploration of social conditions stretching back through Henry Mayhew, Friedrich Engels and William Cobbett to Daniel Defoe, it was not until the appearance of Charles Booth's *Life and Labour of the London Poor*, the publishing of which began in 1889 and culminated in the seventeen volume edition of 1903, that systematic sociological study began.

Significantly, it was the great depression of the early 1870s that marked the opening of the long twentieth century, which prompted Booth, recovering from a physical breakdown caused by overwork in running the Booth Steamship Company, to start to think about the ways in which 'the people' lived. Eventually, he evolved a technique of applying the quantitative methods of business to information gathered from School Board visitors to London Schools (elementary education had been universal since 1870). Later Booth and his researchers devised direct research methods that allowed them to expand their study beyond households with school age children. Ultimately, they were able to show that 30% of London's population lived in poverty: a much higher figure than expected. Subsequently, B. Seebohm Rowntree conducted a survey in York, beginning in 1899 and published in 1901, that relied on direct information from every working-class household and applied a rigid 'poverty line'. Rowntree found 27.84% of the population forced to subsist at a lower level, thus confirming Booth's figure for poverty levels across the country and establishing conclusively that poverty was not the consequence of laziness but a cyclical feature of hard-working, decent working-class lives. The findings of these two studies gave rise to the major Edwardian social reforms discussed above. In 1912, Arthur Bowley of the LSE pioneered a sampling method – selecting householders at fixed intervals from a list of all householders in a town – that produced comparable results to Rowntree and Booth and, by being much cheaper and easier to carry out than those comprehensive surveys, opened the way for the widespread use of such surveys by social institutions.

Although it was strongly challenged by M-O as we shall see, this social survey tradition was to remain dominant within Britain well into the postwar period. It has a number of distinctive features that have peculiarly conditioned the development of the social sphere and Welfare State in Britain. The strengths and weaknesses of the tradition are inherent in Booth's leading declaration: 'That every social problem, as ordinarily put, must be broken up to be solved or even to be adequately stated' (cited in Keating 1976: 24). On the one hand, this emphasis on 'adequately stating' social problems resembles Simmel's insistence that understanding social situations is paramount, but, at the same time, it reveals a fundamentally different outlook: one which ignores the fact that society was in the process of transforming into hitherto unknown configurations and sees only practical problems to be 'solved'. This tendency was compounded by Rowntree's decision to divide poverty into two types: primary poverty – cases where there was simply not enough money coming in – and secondary poverty – cases where, although enough was being earned, money was being diverted into areas considered non-essential so that there was not enough to cover food and clothing requirements. Implicit to this position is a privileging of 'natural' values over those of the social sphere which takes no account of the social necessity of a number of goods and commodities that are not strictly required for biological survival.

Consequently, a characteristically British combination of empirical survey and reactive social policy developed that dealt relatively effectively with the problem of primary poverty without approaching a consistent understanding of wider social dynamics. It is not that the necessary interpretative analytical skills for the latter did not exist in Britain, but rather that they tended to remain confined to literary and cultural realms. Thus, to gain an understanding of how the turn-of-the-century Londoner experiences everyday life in the modern world, one should not turn to Booth but to a commentator such as Ford Madox Ford, who in *The Soul of London* describes how: 'Daily details will have merged as it were into his bodily functions, and will have ceased to distract his attention' (Ford 2003: 10). While Ford remains ambivalent about the mass democracy which has lost its 'fear of [bourgeois] public opinion' (93, 94), he recognises it as something new which, despite its raw, naked, dystopian tendencies, offers a 'Future' if people can adapt to it (99–105). Among those sections of society not adapting, Ford highlights the development of a hereditary poor-but-respectable working class through the situation of a female matchmaker (55–9). The combination of the intense work with the desire to keep her children, both boys and girls, from 'bad ways' led to the children completing elementary schooling only to be kept in with their mother and become matchmakers in their own rights. It was precisely this determination to be 'respectable', which is of course neither economic nor biological but a psychological response to the development of the social sphere and its manifold pleasures, which helped to perpetuate the poverty cycle that the social survey tradition saw in strictly economic terms. Arguably, British social reform served to accentuate respectable working-class poverty by providing only enough support to maintain the cycle and never a sufficient amount to meet the repressed social needs that would have helped people to break out of it.

For this reason, poverty remained a constant presence in the age of social reform even as working-class wages rose and primary poverty declined. Explanations were sought for this unexpected state of affairs and, Booth and Rowntree having ruled out idleness, attention was turned to so-called genetic factors. For example, the Merseyside Survey started by the University of Liverpool in 1929 partially sought to explain persistently high unemployment and poverty rates by the occurrence of 'subnormality' in larger than average families. It was argued, therefore, that the situations of such 'problem families' could not be ameliorated by social reforms. In Mark Abrams's postwar discussion of the history of social surveys, he feels it necessary to point out that the Merseyside results were 'entirely consistent with the conclusions reached by Sir Cyril Burt': namely, that there was 'a negative correlation between innate intelligence and size of family' and 'that the average level of intelligence among the general population may be declining at a rate which might produce serious cumulative effects if at all sustained' (Abrams 1951: 50). Such eugenicist views were not marginal but central to the nascent British Welfare State and can even be found repeated in the report of the 1950 Royal Commission on Population (51). While Abrams, to his credit, provides alternative evidence concerning the effects of nutrition and environment on the results of intelligence testing, he still concludes that the Merseyside Survey 'opened up a new insight into a social problem which had tended to be obscured by the standard social survey into poverty and overcrowding' by facing the possibility that the general economic prosperity supported by social welfare policies might increase certain 'forms of social ill-health' (52). This was the context which made socially acceptable views such as those expressed by the celebrated wartime M-O diarist Nella Last on Sunday 19 January 1941: 'I never thought I'd admire anything Hitler did, but today when I read in the "Sunday

Express" that he "painlessly gassed" some thousands of lunatics, I did so' (Last 1983: 103).

Such attitudes were a direct result of the refusal to differentiate the particular social needs of the long twentieth century from an essentially Victorian conception of poverty. The particular problem, as becomes apparent when comparing the work of Booth and Ford, is that the social surveyors were concentrating on the wrong section of society. If they wanted to understand social dynamics they should have been looking at the emerging mass class of clerks and shop assistants. Ford, albeit ambivalently, saw in the mass parades of such employees before shop windows: 'the great London of the future, the London that matters to the democrat, in the making' (86). However, he also saw this new mass attitude to leisure as a form of social imitation of the upper classes. This wide-held view is refuted by Habermas in his account of the development of a mass culture-consuming public from the smaller bourgeois culture-debating public, which accompanied the structural transformation of the public sphere. Arguing backwards from his time of writing in around 1960, Habermas points out that the distribution of culture-consumption in the forms of magazines, television and radio simply does not support the thesis that a succession of new strata had formed round the nucleus of the old urban-bourgeois reading public. He suggests that it is possible to extrapolate back from the expansion of the new mass public caused by the introduction of television in the United States, where empirical social research shows that the first social groups to buy television sets were predominantly those whose incomes outstripped their formal education:

> If a generalisation be permitted, the consumer strata first penetrated by the new form of mass culture belonged neither to the established stratum of educated persons nor to the lower social strata but often to upwardly mobile groups whose status was still in need of cultural legitimation. Introduced by this trigger group the new medium than spread within the higher social stratum, gradually taking over the lower status groups last (173–4).

These upwardly mobile clerks and shopping assistants are defined by sociologists, somewhat misleadingly, as the 'new middle class' and, as might be expected, were first formally identified in 1890s Germany (Burris 1995: 25). The existence of this class was first put forward in opposition to the then Marxist thinking that all wage and salary earners were part of the proletariat and it quickly generated its own field of social research. While one advocate, Gustav Schmoller, even saw the class replacing the proletariat as the true subjects of history, the conventional viewpoint was that they would develop into a stabilising mediator between labour and capital. This attracted the revisionist wing of the Social Democratic Party, with Emil Lederer identifying salaried technicians and commercial employees as occupying a new class position in a work published in 1912. However, economic decline during and after the war, led Lederer to alter his position. In 1926, writing with Jacob Marschak, he predicted that an intermediate class position would become untenable, forcing the new mass of salaried employees to side with either the forces of capital or labour (see Burris 1995: 25–9). After considering the unionisation of these salaried employees, they concluded that 'a single stratum of all gainfully employed [i.e. waged and salaried] ... is in the process of formation' (Lederer and Marschak 1995: 82). This position was challenged by Hans Speier, writing in 1932: '... of the three types of unions that white-collar workers could have joined during the Weimar years only the "middle-class" organisations *gained* in size and importance. Towards the end of the Weimar Republic they emphasised the middle-class self-understanding of white-collar workers even more sharply than they had at its beginning' (Speier 1995: 97) – a line of argument that anticipates the later ascription of the blame for the rise of National Socialism to the 'status panic' of white-collar employees (see Burris 1995: 41–6).

However, before that historical point had been reached, a former student of Simmel, taking Lederer as his point of departure, launched an interpretative study of the new middle class within the wider context of everyday life. Siegfried Kracauer's *Die Angestellten*, first published as a serial in the *feuilleton* of the *Frankfurter Zeitung* in 1929 anticipates Speier by rejecting Lederer's optimism with regard to the 'new middle class' adopting a proletarian outlook, but argues that far from the salaried employees increasingly coming to identify with their employers, they were becoming increasingly alienated precisely because they could not identify with the 'automatic course of free competition' that management substituted for any idea of human order:

> The blame for this is hard to apportion, and at any rate lies only partly with the employers themselves. During the post-war period they not merely had to find their way in altered social and eco-

nomic conditions, they were also saddled with the demand that they fill the vacuum left behind by the vanished former upper class ... They try to master it by transforming the old form of rule into an enlightened despotism that makes concessions to the socialist counter-current .... All the compromises prove only, of course, that for the sake of the sovereign economy the employers are adapting to present conditions – but without basing themselves upon them. A stratum thus finds itself in power which, in the interest of power and at the same time against this interest, cannot found its own position ideologically. But if it shrinks itself from confronting the reason for its existence, the everyday life of the employees is more than ever abandoned (Kracauer 1998: 100).

This change in the position of management had of course already commenced well before the First World War as we know from Habermas's analysis, but what Kracauer highlights here is a gap between twentieth-century socio-economic functions and nineteenth-century cultural representations. A crisis of self-confidence following defeat in the war and the abortive German Revolution left both employers and employees with only the outmoded bourgeois values of the old century to orientate themselves by, which, of course, did make the new middle classes look imitative. If this was uncomfortable for the employers, it was unbearable for the employees caught within what Inka Mülder-Bach describes, in the introduction to the English translation of *Die Angestellten*, *The Salaried Masses*, as 'the tension between proletarianised existence and bourgeois self-definition' (Mülder-Bach 1998, 6). However, the point is not that this tension would eventually drive the new middle class into the arms of the Nazis but that it masked the real possibilities of social transformation in a situation where employees 'for the first time became the formative power of the public sphere' (5). The real innovation of Kracauer's investigation of these salaried employees is the deployment of cinematic montage and close-up techniques, which allow him to combine ethnographic study with an implicit promise of social transformation suggested by the juxtaposition of images. Thus, his work can be seen both as a forerunner to M-O – albeit unknown to them – and more importantly as part of a shared interwar European cultural politics that, although increasingly acknowledged, is still not widely understood in terms of its scope and significance.

In practice, Kracauer links cultural analysis and personal observation to existing sociological and statistical analysis to investigate facts such as 'that employees do indeed devote less money to food than the average worker, but they rate so-called cultural needs more highly' (Kracauer 1998: 89). In identifying popular culture as a culture of distraction – prefiguring Adorno's account of the 'Culture Industry' – Kracauer explains it as a manifestation of the social need for distraction from the default condition of everyday life: 'an everyday existence outlined by the advertisements in magazines for employees. These mainly concern: pens; kohinor pencils; haemorrhoids; hair loss; crêpe soles; white teeth; rejuvenation elixirs; selling coffee to friends; dictaphones; writer's cramp; trembling, especially in the presence of others; quality pianos on weekly instalments; and so on.' (88). The possible supersession of this pursuit of distraction is suggested by the unceasing tendency of the kaleidoscope of cultural fragmentation to form into new patterns: providing fleeting snapshots of a future world in which respectable proletarian-born office girls and salaried bohemians co-exist in harmony. Kracauer, of course, never confuses this work of storytelling with emancipation itself and allows himself just one moment of lasting illumination in describing the life of a cigarette salesman, like a prince in his company car and yet 'unaffected by the high life', who, feeling at home with his union colleagues in a coaching inn, begins to sing: 'Then everyday reality vanishes and all those present, far beyond the circle of colleagues, revel in the enjoyment of a more beautiful life' (69). A paragraph break allows us to pause before the next line: 'Not often does the economy leave open such a gap, in which a person from a lower stratum who is something is permitted to be just as he is'.

As Mülder-Bach rightly observes (4–5), Kracauer's approach resists the 'methodological generalisation' of the similar participant observation method employed by the Robert and Helen Lynd in *Middletown: A Study in American Culture*, published in the same year as *Die Angestellten* and the book which inspired M-O to rename Bolton as Worktown. However, while *Middletown* does not display the same critical stance as *Die Angestellten*, it does register similar changes in the social and cultural spheres, especially the commodification of spare time: 'leisure time ... quite characteristically in a pecuniary society, is "spent"' (225). The book represents a landmark in the history of social surveys for relating consumption to social psychology and linking this relationship to the historical development of the society studied. Therefore, not only is the rise of social – as opposed to biological or moral – necessity since 1890 charted through such fields as changing styles in women's clothing and rates of automobile ownership, but also this social necessity, itself, is revealed as the motor of social transformation:

> The rise of large-scale advertising, popular magazines, movies, radio, and other channels of increased cultural diffusion from without are rapidly changing habits of thought as to what things are essential to living and multiplying optional occasions for spending money. Installment buying, which turns wishes into horses overnight, and the heavy increase in the number of children receiving higher education, with its occasions for breaking with home traditions, are facilitating this rise to new standards of living. In 1890 Middletown appears to have lived on a series of plateaus as regards standards of living; old citizens say there was more contentment with relative arrival; it was a common thing to hear a remark that so and so 'is pretty good for people in our circumstances.' Today the edges of the plateaus have been shaved off, and every one lives on a slope from any point of which desirable things belonging to people all the way to the top are in view (81–3).

This depth and focus of study were produced by the use of a sixfold classification, taken from the anthropologist W.H.R. Rivers, under which the activities of any culture could be described – apparently first selected as a suitable frame of reference for the original incarnation of the project as a survey of religious provision and practice in a typical small city (see J. Madge 1963: 126–7). Although the application of anthropological approaches to domestic society had been pioneered by the Chicago School with particular emphasis on the investigation of deviancy and social problems, the Lynds produced a picture of society as a whole, in which the term social problem appears only infrequently and then often in inverted commas. The success of the book, which ran to six printings in the first year and remains in print today, is evidence that this approach not only generated a study of the general public but also a study that was for the general public. However, a major consequence of this anthropological and historical approach – described by the Lynds as 'dynamic, functional study' (6) – was the binary division of Middletown society into a 'Working Class' and a 'Business Class': 'Members of the first group, by and large, address their activities in getting their living primarily to *things*, utilizing material tools in the making of things and the performance of services, while the members of the second group address their activities predominantly to *people* in the selling or promotion of things, services and ideas' (22). This unusual and ultra functional division affects the way that the social data is interpreted. Collapsing everyone from clerks to the wealthiest family (revealed by the 1935 sequel *Middletown in Transition* to be running the key industries, the Chamber of Commerce and the local political machinery) into one class obscured the power relations of the city, but allowed the Lynds to highlight how all are 'merged in the life of the mass of business folk' (23n). This identity, they suggest, was partly negative: 'As the study progresses, the tendency of this sensitive institution of credit to serve as a repressive agent tending to standardize widening sectors of the habits of the business class – to vote the Republican ticket, to adopt golf as their recreation, and to refrain from "queer," i.e., atypical behaviour – will be noted' (47). However, elsewhere, as in the long quote above, credit is seen as a means of satisfying social needs and telescoping 'the future into the present', not just for the Business Class but for every family in Middletown (46).

What *Middletown* shows implicitly is how the 'new middle class' or, to use its own terms, the new recruits to the Business Class, cemented their gains in social status by acts of consumption which forced other more established families to follow suit. The immediate effect might have been standardisation, but it also served to flatten social hierarchy. While, as the appendix describing research methods makes clear, the ruling family were omitted in practice from the first study (510), *Middletown in Transition* shows how much of their activity was directed not only to economic and political matters but also to the wider social sphere of Middletown, especially health, education and leisure provision. Critics have remarked that the Lynds do not seem particularly opposed to this family's apparent hegemonic rule (see J. Madge 1963: 155) and the reason might well be that they recognised these developments as driven primarily by the social needs of the wider Business Class or, indeed, as demonstrating what Kracauer found to be missing in Weimar Germany: a leadership capable of finding its own position ideologically. The activities of that particular Middletown family can thus be seen as a particular manifestation of the wider process that was subsequently identified by Habermas, in which new cultural and consumer values spread from the socially mobile across the highest social stratum. In further accordance with the Habermas model, the same values can also be seen to have taken over the lower status groups in 1935, so that the Lynds were satisfied that the Middletown Working Class no longer subscribed to a separate class identity but adhered to the same consumer lifestyle (J. Madge 1963: 158). Thus the fascination of reading *Middletown* today is finding the widespread everyday values and concerns of modern Western society, from washing-machines to fourteen year-old daughters wanting to stay out until 11.00 pm, already present in a small American city in 1924: 'However drab or shadowed by fear [of marital breakdown or failure to meet payments] these homes may be, there are always the plans for today and tomorrow, the pleasures of this half-hour, the "small duties and automatic responses to the custom of the daily round of living which imperceptibly but surely mitigate the tragedies and disappointments of existence"' (130).

In conclusion, it can be seen that forms of social survey had been developed by the end of the 1920s that met Simmel's demand for means of understanding the possibilities raised by the structural transformations of the public sphere which had characterised the opening decades of the long twentieth century. M-O's unique contribution to this tradition would be to combine modernist techniques holding open the possibility of social transformation, such as those deployed by Kracauer, with an anthropological framework, such as that used by the Lynds. By recruiting a self-selecting 'new middle class' membership, M-O was able both to highlight emergent developments in mass society and to help the 'new middle class' wrestle the means of cultural reproduction from older social strata; with the result that British society was fundamentally transformed, albeit not necessarily in the directions initially intended. However, before we turn to this project in detail, it is necessary to consider the formative social and cultural influences on the M-O founders and the specific traditions and movements to which they reacted.

2
The Space of Former Heaven

Cambridge *Experiment*

Although Harrisson, Jennings and Madge were all public-school educated and went to Cambridge – where only Jennings actually obtained a
degree – it seems appropriate to follow Orwell's example and describe
them as members of the lower-upper-middle-class (Orwell 1998: 113)
because none of them had the means to live the life they were theoretically brought up for. Like Orwell and their other contemporaries of the
Auden generation, the M-O founders are open to the accusation that
their interests in the masses, and the working class especially, were
mainly a form of 'negative identification' in which the need to reject
the outmoded social values of their bourgeois upbringings outweighed
any real commitment to the social experience of the group selected.
Valentine Cunningham has stated this position particularly baldly: 'The
cult of the worker for such worshippers was never far from the search for
the lost authoritative father-figure (Madge's father, Colonel of the Royal
Warwickshire Regiment, had been killed … in the war)' (Cunningham
1988: 242). However, the important criterion for making such judge-
ments is surely not the initial identification with the social other, which
was probably negative of necessity, but the extent to which this identification was subsequently transcended allowing the bourgeois intellectual to progress beyond satisfying individual needs to attempting to
satisfy social needs as a whole. It is in this context, as demonstrated
across this and the following chapter, that M-O can be seen as a key test
case for assessing radical European cultural politics in the 1930s.

A different, but complementary, perspective is suggested by Raymond
Williams's discussion of Orwell's negative identification in terms of
national identity rather than social class. Williams argues that, because

he was born in India and spent most of his childhood in educational institutions of his own social class before going to Burma for five years as an imperial policeman, once Orwell rejected the values of the colonial social system, he had no organic point of contact with England and could only find one by the negative process of becoming a down and out: 'there was no other "England" to which he could immediately go. He could only drop out of the one England and make expeditions to the other' (Williams 1991: 19). This enabled Orwell to develop, as the second phase of his relationship with England, an acute sense of social contradictions. The relationship entered a third phase when, according to Williams, under the pressure of the war years, Orwell created 'a new myth ... of an England of basic ordinariness and decency' (22). This model can readily be applied to the history of M-O and particularly to the life trajectories of Harrisson and Madge, who had the same imperial and educational background as Orwell, being born respectively in Argentina and South Africa and educated respectively at Harrow and Winchester. Not only did Harrisson tramp like Orwell, but his eventual plunge into the industrial working-class life of Bolton followed immediately after his two year sojourn in the New Hebrides, which had included a spell as a colonial official. Madge grew up, and arrived at university, a self-declared imperialist before his 1932 conversion to communism took him to the backstreets of Cambridge in regular expeditions selling the *Daily Worker*. In a poem of the following year, 'In Sua Voluntade', he expresses the general feeling of exiles from the imperial classes: 'England is fallen. Home is gone. Time stands' (Madge 1994: 17). In this absence, Madge and Harrisson's later joint work, *Britain by Mass-Observation*, can be seen like Orwell's *The Lion and the Unicorn* as the attempt to create a new national myth. The history of both M-O and Orwell can be seen as the outcome of the need to imagine a national identity for the 'home' country because of neither belonging by birth nor being able to accept the economic and political ties that connected their birth places to that 'home' country. The process is a near reversal of that described by Benedict Anderson in his account of how the idea of nationalism originated with the Creole populations of the Americas, who were not indigenous but wished to be no longer tied to the centres of empire from which they had originated (see Anderson 1991: 47–65; Hubble 2004: 29–41). The question of evaluation hinges on whether one follows Williams's line that negative identification leads to the creation of a necessarily false myth or whether one sees M-O as imagining communities that transcend self-interest.

Being at Cambridge was undoubtedly an advantage for intellectually understanding the scope of the structural transformation of English society and the widespread changes in social needs. Such concerns formed the rallying cry for 'Cambridge English' as seen for example in the opening page of I.A. Richard's *Science and Poetry*:

The fairly near future is likely to see an almost complete reorganisation of our lives, in their intimate aspects as much as in their public. Man himself is changing together with his circumstances; he has changed in the past, it is true, but never perhaps so swiftly. His circumstances are not known to have ever changed so much or so suddenly before, with psychological as well as with economic, social and political dangers (Richards 1926: 1).

Richards's argument, of course, was that 'poetry is capable of saving us; it is a perfectly possible means of overcoming chaos' (82–3). As Terry Eagleton has argued, Richards, the son of a works manager in Cheshire, represented (along with others such as F.R. Leavis) a new social class – we might want to be more specific and label it the new middle class – entering the traditional universities and the subsequent development of 'Cambridge English' was tantamount to a revolution in English studies, which remains unsurpassed for its 'courage and radicalism' (Eagleton 1983: 30–1). While the influence of Leavis's journal *Scrutiny* and his moral emphasis on 'Life' is well known and well documented (see Mulhern 1981), it was Richards, emphasising poetry's virtue as being precisely that it can be disentangled from moral belief, who sometimes had to lecture in the street because the hall was not big enough (Jackson 2004: 59). The amazing sequence of books Richards wrote in the 1920s, including *Science and Poetry*, *Principles of Literary Criticism* and *Practical Criticism*, retain their place in the history of literary criticism without appearing to have had any lasting effect on society at large. Yet given the close relationships that Madge, Jennings and their sometimes M-O collaborator and friend, William Empson, all had with Richards while at Cambridge, it is possible to argue that M-O represents the social legacy of Richards's thought.

It is true that in a 1979 interview with Angus Calder, Madge played down the idea of Richards having any great significance for M-O, but when pressed on Richards's use of 'ordinary' poetry readers in *Practical Criticism*, he admitted 'maybe it did more than I realised' (Madge 1979: 3). Moreover, the fact that the original working title for

M-O was PP suggests that Richards was a direct influence – a link which will be further explored in the discussion in the third section of this chapter on the wider importance of poetry to the English Left in the 1930s. In the meantime, other connections can be traced through the intermediate form of the journal *Experiment* – a name which reiterates the link between science and poetry – which Jennings and Empson helped to found at Cambridge.

Jennings, the subject of a recent major biography by Kevin Jackson, was born and grew up in Walberswick, a rural village on the Suffolk coast. His father, Frank, an architect, and his mother, Mildred, a Slade-trained painter, were Guild socialists and friends of A.R. Orage, editor of *The New Age*. On Orage's recommendation, Jennings was sent to the Perse School, Cambridge, in 1916. Where, after a decade he was 'equipped with an exceptionally fine training in modern languages and classics, unconventionally advanced skills in English composition and a vivid and well-formed historical sense. He had shone as an athlete, an orator, an actor and designer, and had published his first poems' (Jackson 2004: 45). In the Autumn of 1926, Jennings arrived at Pembroke College, Cambridge, as part of the first year allowed to study English literature without combining it with a different subject. A brilliant student, he achieved firsts in both parts of the English Tripos and proceeded, now married to Cicely Cooper, to a postgraduate study of the poetry of Thomas Gray under the supervision of Richards. Besides studying, painting, courting and working in the theatrical world – designing full-scale productions at the New Theatre in Cambridge and occasionally acting – Jennings also found time to help found *Experiment* in the academic year 1928–9. This was the first organisation set up by people who were subsequently to feature in M-O: Jennings and Empson sat on the initial editorial board, while Kathleen Raine and Julian Trevelyan were also heavily involved. Another related trajectory is that leading to the English Surrealist Group, of which Jennings and another editorial board member, Hugh Sykes Davies, were later to become founder members. In general, *Experiment* was at the centre of a gifted Cambridge generation – other major participants included Jacob Bronowski and Malcolm Lowry for example.

Although the journal only ran to seven issues, it made a genuine impact at the time as can be seen from the large selection featured in the June 1930 – at which time *Experiment* was being edited by Bronowski and Sykes Davies – edition of Eugene Jolas's *Transition* under the title 'Cambridge Experiment: A Manifesto of Young England'. This began appropriately enough with an actual manifesto, unsigned, which reads in full:

> If we were hawking some sharp and particular quarrel: with Miss M for a charlatan, Mr N for a Christian: if we had discovered some literary panacea: or renounced some popular philosophy: it would be easy now to be pointed, to be vigorous, to be witty. But because we have no panacea: and continue in humility: you will find us quite simply clumsy. We have lost faith, you see, in this tinkering with the structure of literature: and if you find us labouring it is because we are trying to shift, ever so slightly, its bases. Rather accidentally; because we invented no principles; and now that they have happened to us, they are uncertain and not at all startling. A sense that literature is in need of some new *formal* notation: an attempt to show how such a notation can be built out of *academic* notations, where academic means perhaps no more than non-moral and is after all best explained in our poetry: a belief in the compact, *local* unit: and in the *impersonal* unit: a belief finally, and a disbelief – for it is about this mainly that we are at odds – in *literature* as a singular and different experience, something more than an *ordering* of life. You see how haphazard it all is. And its criterion ultimately is only again *Experiment* (106).

This clearly displays the influence of Richards in its references to academic notation equating to a non-moral approach and also suggests a movement beyond Richards in its suggestion of literature being more than an ordering of life. One of the following articles (also unsigned), an extended gloss on the concept of 'Localism' referred to in the manifesto, contains the note: 'perhaps I.A. Richards is in danger of tripping over literary criticism' (116). The rejection of founding principles was later to characterise the early phases of M-O and stems from a common concern of the generation with prematurity, articulated in the notes on 'Localism': 'precocity is the prevalent moral and intellectual disease of our time'. Variants on this theme were to re-emerge later in the decade in examples such as Madge's poem 'Instructions' and 'Proletarian Literature', the opening essay of Empson's *Some Versions of Pastoral*. Indeed this article captures a nodal point in cultural politics, which would be recapitulated by M-O, in that it seems equally to anticipate Empson's arguments and those of the 1934 Soviet Writers' Conference, to which he was part-satirically responding. For while it celebrates localism as the possibility of representing the complex in the simple, it also proclaims '*the failure* of the cosmopolitan art [i.e. modernist art] up to the present'. Furthermore, it nonetheless states optimistically: 'In time there may again arise a possibility of synthesis which will produce a unit larger than any yet known, and whose resulting art should be richer because it contains more, and more digested, material.' There is little doubt that Jennings and Madge would later, at least for a while, see M-O as the vehicle for this richer synthesis.

Jennings's name does not feature in the selection for *Transition* and, indeed, his editorial work at *Experiment* seems to have been short-lived, but he contributed a number of articles to the journal across its lifetime covering his interests in poetry, painting and theatre design. Many of these articles were co-written and one of his co-authors, Gerald Noxon, has testified to the difficulty Jennings had in reducing the flow of his thought to finite text, thus necessitating a collaboration which 'was mostly my writing and his thinking' (cited in Jackson 2004: 110). This pattern would be continued in M-O, in which – given that Harrisson was also a better talker than writer – Madge ended up undertaking most of the drafting of ostensibly co-authored texts. However, in a wider sense, this generation's ideas were developed collectively such that Jennings's 'Notes on Marvell's "To His Coy Mistress" suggests, according to Jackson (71), a way of thinking about poetry similar to that which Empson was shortly to reveal in *Seven Types of Ambiguity*, and which was developed in wide discussions between the two (107–9). Indeed the inevitable focus of literary criticism on written texts not only does Jennings a huge disfavour, but also the Cambridge generation as a whole. Focusing on Jennings's other cultural activities shows how *Experiment* was really a movement. During the same period at the turn of the decade, Jennings established the Experiment Gallery, reflecting the latest movements in painting including work by himself and Trevelyan (100–1); published an edition of Shakespeare's *Venus and Adonis* for the Experiment Press (106); and planned an Experiment theatrical production of Marlowe's *Dido* (104–5). This integrated cultural approach prefigures both Jennings's conception of M-O and the wartime films for which he is chiefly remembered. It also suggests that just as *Experiment* needs to be considered as more than a collection of written texts, so likewise does M-O.

In 1932, while convalescing from intestinal tuberculosis, Madge, who had arrived at Magdalene – Richards's college – the previous autumn to read Natural Sciences, read Empson's review of Auden's *Paid on Both Sides* in the final Spring 1931 edition of *Experiment* and then read Auden himself. The experience of this illness informs the first two stanzas of a poem written later that year, 'Letter to the Intelligentsia'. The third begins:

> But there waited for me in the summer morning,
> Auden, fiercely. I read, shuddered and knew
> And all the world's stationary things
> In silence moved to take up new positions (Madge 1994: 130)

This epiphanic moment, which acted as a catalyst on Madge's poetic career as discussed in the third section of this chapter, was shortly compounded by another rapid conversion that started a very different Cambridge-based trajectory leading to M-O. Having seen NUWM 'hunger marchers' arrive to bivouac near Magdalene, Madge went to talk to them and as a result joined the university branch of the CP. Writing in his unpublished Autobiography, Madge went on to record how 'on Sundays we used to sell *Daily Workers* in the back streets of Cambridge, going from house to house often being asked inside. This was, of course, instructive not so much politically as sociologically, and provided a background of sounds and sights and smells against which "the working class" appeared more concretely' (35–7). Thus, a lifelong interest in working-class homes was started which continued through M-O and his postwar spell as Social Development Officer at Stevenage New Town up to his work in the late 1970s on inner city poverty with the Institute of Community Studies.

Madge's immediate political career at Cambridge was extremely productive, given that he only spent another year there, encompassing the founding of the journal *Cambridge Left* and being elected to succeed David Guest as CP organiser. However, John Cornford ended up assuming that position because Madge eloped with Kathleen Raine, by then the wife of Sykes Davies. Or as Raine records it in her autobiography:

> I allowed myself to be rescued from my first marriage by Charles Madge; for this I was altogether to blame, for I was older than he, and I allowed him, in the chivalry of his poetic vision of me, to wreck his University career for my sake ... What he saw in me, only God knows ... perhaps his poetic muse personified, since he knew me to be a poet ... perhaps, even, a woman of the people, whom as a Communist (as he had just become at that time) he could fitly idealize... I remember well the terms in which he offered himself to me: 'Come with me,' he said, 'and I will give you a cause to live for.' What he offered me was the cause of Communism ... (Raine 1975: 78).

Their first child was born in June 1934 round about the same time as Madge was receiving the commendations of CP General Secretary Harry Pollitt for *Fight War and Fascism*, a monthly newspaper that he was editing with Guy Hunter, a fellow Wykehamist (i.e. former pupil at Winchester) who he had persuaded to follow him into the CP at Cambridge. The organisational and editorial experience Madge gained with the CP was invaluable for the foundation of M-O, in which Hunter was also an early participant.

Madge noted in his Autobiography that his embracing of communism represented 'a very rapid and complete change of orientation' (37). On one level this was completely true because, as already noted, at the time of his arrival at Cambridge he considered himself an imperialist. However, in his interview with Calder, Madge qualified this with the information that 'My father in South Africa had been part of Milner's Kindergarten – and I formed a kind of admiration towards the constructive imperialist activity' (Madge 1979: 26). Lord Milner had led the reconstruction of South Africa following the Boer War and his 'Kindergarten' was the administrative force responsible for scientifically planned progress. From the evidence of clippings in the Madge papers, Madge's father, having fought in the war, seems to have worked particularly in the area of Land Settlement, being involved in various committees and even visiting England some years prior to Madge's birth to lecture on 'Immigration and Settlement of Whites on the Land' (CMP: 1/3). This plan of Milner's had been initially intended to promote British immigration and dilute the influence of the Boers, but white immigration remained a priority even after the establishment of the Union of South Africa in 1910. Colonel Madge went on to become Land Manager of the Transvaal Consolidated Land Company and Charles was born on 10 October 1912 in Johannesburg with brother John following on 19 June 1914 (Madge 1987: 1–2). After rejoining the army at the outbreak of the First World War, Colonel Madge first worked as Director of the Information Bureau at Defence Headquarters in South Africa. A letter from General Smuts notes that 'Colonel Madge has done exceedingly good work, which is none the less meritorious because it has been of somewhat dull and prosaic nature. It has, however, meant the constant exercise of no small organising ability' (CMP: 1/3). His son would later decline the opportunity to work for the MOI during the Second World War, but his work instead with Keynes, Beveridge and PEP played a role in postwar reconstruction that recapitulated his father's efforts, albeit for rather more democratic ends. Viewed in this context, the switch from constructive imperialist to communist was not such a complete change of orientation.

In September 1915 the family returned to England, in order for Colonel Madge to play a fuller role in the war. When he finally arrived in Boulogne on 6 May 1916, however, it was only to die four days later after being struck on the head by shrapnel. The pathos of the circumstances was accentuated by his last letter, dated 9 May, in which he wrote that while he did not really mind not having a steel helmet, 'the only trouble is that you can't go out for a good walk' (CMP: 1/3). Not surprisingly, Madge grew up close to his mother, Barbara, who began regularly to transcribe his poetry. By way of return, Madge produced a selection of her letters in 1926 in an exercise book, complete with preface: 'They were written day by day, and so relate to daily tasks and enjoyments. Many of the extracts are as polished as gems but all have the fragile charm of woodland and downland flower, and the timbre of birdsong' (CMP: 8/1). Over his last three years at Winchester, Madge carried out an active reading programme including, on the one hand, Richards, Empson, T.S. Eliot and Ezra Pound and, on the other, works such as Sir James Frazer's *The Golden Bough*, Bronislaw Malinowski's *Sex and Repression in Savage Society* and W.H.R. Rivers's *Psychology and Ethnology*. As well as this, and winning a scholarship to Magdalene, he also wrote a book intended for inclusion in the Kegan Paul series 'Books of the Future' entitled *Arethusa or the Future of Enthusiasm* which dealt with 'applied psychology in the context of industrial society' (see Madge 1987: 19–30). Therefore, it can be seen that even before he reached Cambridge, Madge's education had already fitted him for the uniquely varied career he was to follow. One, as Calder noted in his obituary of Madge, that reflected through a biography would 'cast light on the highways of twentieth century life, as well as in many interesting corners' (Calder 1996).

Letter to Oxford

Harrisson was born on 26 September 1911 in Argentina, where his father worked as a railway engineer. A Boer War veteran, like Colonel Madge, Harrisson's father also brought his family back to England at the outbreak of the First World War so that he could join the army. Unlike Colonel Madge, he came through the war to emerge a much-decorated Brigadier-General before returning to Argentina and the railways. Harrisson and his brother were left behind at boarding school, but they did spend a year with their parents between the summers of

1922 and 1923. By now General Harrisson was running an Argentinean railway network and so the family lived in high colonial style: riding every morning, shooting, hunting, tennis and travelling widely in two specially fitted railway coaches around the network. But Harrisson, grandson of the naturalist William Eagle Cole, also developed a growing interest in birds. This was to become an important pastime when he started at Harrow in 1925 and ultimately the source of his subsequent career (Green 1970: 102–4; Heimann 1998: 1, 9–13).

The year in Argentina made Harrisson acutely conscious of 'a feeling of belonging to England and *not* belonging to it' (cited in Heimann 1998: 12), which was presumably reinforced by the seemingly endless cycling of educational institutions, colonial adventures and spells of dropping out which comprised his subsequent life story. Whether this suggests a succession of negative identifications or an identification only with this state of duality itself is open to question. There is no doubt, though, that Harrisson chose to present it as an unqualified advantage with respect to understanding his 'home' country: 'If you are not born and brought up in England it gives you a much more objective attitude to the country when you arrive' (cited in Green 1970: 102). However, it could equally be considered a coping mechanism. For example, one way that Harrisson dealt with being at Harrow was to devise a classification system for his fellow pupils by filling out a filing card for each, which enabled him to identify everyone in the school.

This also seems to bear out a claim, frequently directed at Harrisson's later career and work with M-O in particular, that 'his anthropology took on a particularly positivist character: it could be seen as treating human beings like birds, to be observed and described as though they were natural specimens' (Sheridan et al 2000: 83). Yet this is to confuse the use of classification for identification purposes, which tends to be positivist by definition or it is not a great deal of use, with the entire purpose of ornithology, which is a branch of natural history. Harrisson's first publication, 'Birds of the Harrow District 1925–30', compiled while at school, illustrates why ornithology does not reduce to empirical observation, but relies on a complex interpretative stance. Thus, Harrisson combines his quantitative analysis, that bird population in Harrow is increasing 'based on actual census work, carried out annually', with a qualitative analysis which takes into account synchronic as well as diachronic factors: 'I venture to suggest, on the evidence so far obtained, and as a tentative theory ... that this is due to an increased concentration of birds in the undeveloped areas of Harrow – faced by the development of houses and housing estates in the surrounding area .... Thus the "increases" observed by ornithologists in this district are not true increases, but apparent increases' (Harrisson 1930: 84–5). It could therefore be argued from Harrison's career that ornithology is a very good basis for a discipline such as anthropology precisely because it offers a model for the effective combination of positivist and interpretative analysis. For example in a wartime article for *Sociological Review*, Harrisson returned to his Harrow experiences to illustrate a point about social hierarchies:

> ... there was an elaborate system of privilege and caste, mainly based on the length of time you had been there, but also on how good you were at games. The speed at which you could move downstairs, which waistcoat buttons could be undone, which hand you could put in your pocket, what cereals could be eaten at breakfast, where you could walk, a hundred habits, were determined entirely in this way (Harrisson 1942b: 152).

It seems likely that it was Harrisson's file-based classification key that allowed him to identify pupils in the first place and thus made it possible for him to interpret correctly the many everyday scenes played out before him and so generate this understanding of the social system as a whole. It can further be argued that there was no difference between the practical strategy the young Harrisson employed to adapt to the strange society he found himself in – a strategy he was to repeat throughout his lifetime with each new strange society he encountered – and the manner in which he consciously understood that society. Parallels can be seen with the *Verstehen* tradition, in which the interpretative understanding of social life is supposed to be based on a developed version of everyday understanding, but in Harrison's case it looks as though he did not so much intellectually choose to develop his everyday understanding as that he was inescapably conscious of it because of the duality arising from his colonial background. Inversely, the great world historical problem that Simmel had identified concerning the two ways of defining the position of the individual to the totality was something that Harrisson experienced at an everyday level. As a result, he could neither accept a purely intellectual or everyday practical approach and much of his early life was driven by the search for a satisfactory combination of the two.

This search made his time at Cambridge into a deeply unsatisfactory experience. In the autumn of 1930 he was admitted to Pembroke College but whereas Jennings, three years before him, had reacted to the 'hearty' sporting character of that college by simply allowing his many cultural and intellectual pursuits to lead him into a wider sphere, Harrisson loathed it because he wanted to be 'tough' and 'intellectual' at the same time (Heimann 1998: 20–1). Therefore, he spent his time on running, swimming, cycling, ornithology and drinking. Interestingly, it was through the latter activity that he made his only contact with members of the *Experiment* circle: he gatecrashed a party and met Lowry and John Davenport, who became drinking friends. Lowry apparently met Harrisson's intellectual-but-tough criteria by virtue of having spent six months as a merchant seaman. Nevertheless, Harrisson left Cambridge after the first term of his second year and so caused a rift with his father which would culminate in his disinheritance in 1937. However, he only moved as far as Oxford where he took up residence early in 1932 in the Balliol rooms of an old schoolfriend, Reynold Bray – briefly to be involved with M-O in 1937 in Bolton and Blackpool before an untimely death in the Arctic. Harrisson was there because he was organising for Oxford University's four-month scientific expedition to Sarawak, which he was to lead during the summer. He was already the veteran of two scientific expeditions – for which he had been selected for his proficiency in ornithological fieldwork – in the summers of 1930 and 1931 respectively and by the time he had returned from Sarawak, preparations were already underway for the 1933 expedition to the New Hebrides that would eventually keep him away from Britain for more than two years. During the six months between his third and fourth expeditions, Harrisson lived a bohemian lifestyle, bearded and longhaired from the Sarawak expedition, wandering around Oxford in sandals with his toe nails painted red. Through John Baker, the zoologist in charge of the New Hebrides exhibition, he made a number of friendships with women who would subsequently play significant roles in the history of M-O (Heimann 1998: 38–9). One of these was Mary Adams, a former research student into cytology, who was in charge of adult education for BBC radio. Adams went on to become the first female television producer for the BBC when the service was introduced in 1936, thus facilitating a number of television appearances for Harrisson both as a popular anthropologist and, later, as a member of M-O. More importantly, she became the director of the HI section of the MOI in 1940 and employed M-O as the key source for reports on domestic morale. The other two friends Harrisson made were Baker's wife, Zita, and her best friend Naomi Mitchison. Mitchison, who was also a friend of Madge and Raine, went on to write a major wartime diary for M-O, published much later under the title *Among You Taking Notes*. Harrisson and Zita Baker were lovers on and off from early 1933 until 1937, when they spent time together in Bolton during the earliest days of the M-O Worktown project. Zita also participated in the August 1937 M-O survey of Blackpool, where, in the interests of investigating holiday sex, she stood alone outside the Tower and was repeatedly pro-positioned. By this time she had left Baker and was married to Dick Crossman but she still kept in touch with M-O, going on to write a long report on the May 1940 Labour Party Conference (Calder and Sheridan 1984: 42, 44, 54, 187–98).

On 9 February 1933 the Oxford Union famously voted for the motion 'That this House will under no circumstances fight for King and Country' and then reaffirmed its decision with a larger majority on 2 March. As Judith Heimann suggests, this appears to have been the catalyst, or at least a factor, in Harrisson sitting down between 28 March and 6 April 1933 to write (or, rather, dictate to Zita) the ninety eight pages of *Letter to Oxford*: 'Tom asked: If the Ox was not prepared to fight for King and Country, what *was* he prepared to fight for, die for, or even risk embarrassment for? The answer seemed to be "very little"' (Heimann 1998: 39–40). It is true that Harrisson was com-plaining about an Oxford fear of engaging in the experiments and enthusiasm of youth, but at the same time he was also trying to work out a dialectical resolution between intellect and action for his own benefit as well as for that of the wider generation. As his argument unfolds it begins to anticipate a number of trends and ideas that would later feature in M-O publications.

The limitation of *Letter to Oxford* lies in the attempt to map the opposed attributes on to Oxford and Cambridge: 'Ox is the home of the mind, Cam of the body' (22). This necessitates the implication that 'Cam' intellectuals such as Empson are excessively highbrow to the point of sterility, which is backed up by a humorous footnote: 'For a pleasant breakfast-book try *Seven Types of Ambiguity* by William Empson. The critics F.R. Leavis and I.A. Richards are typically Cam. Ezra Pound has a Camisoul' (ibid). Further on, another of Harrisson's future colleagues is made fun of in connection with the poetry journal *New Verse*: 'in which Charles Madge is featured every month in *Comic Cuts* of his fraternal poets' (87). Harrisson's strain of anti-intellectualism and anti-aestheticism would occasionally resurface in the future but it is actually less damning than the faint praise he showers on Oxford for its pleasant but timeless sense of unreality which breeds inaction: 'Stephen Spender, the most important recent Oxpoet, has just published his *Poems*. Fifty per cent exhibit this timelessness' (16). Harrisson's future criticisms of Spender and Auden, which became more strident, provoked their supporters in the press and caused savage criticisms of M-O in retaliation – a tendency which has persisted to the present in literary critical accounts of the 1930s.

The strength of the book lies in its diagnosis of the roots of the current inertia affecting Harrisson's generation – a diagnosis which has plenty in common with the 'typically Cam' approaches of Richards and *Experiment* – as lying in the unprecedented social and psychological changes occurring in the twentieth century: 'We refuse to admit that the whole framework of life has been changed – not politically but mentally, aesthetically and spiritually; politics is thirty years behind' (50). However, Harrisson lends this analysis a sense of urgency that was to be one of his characteristic contributions to M-O, by linking it to a timely warning of the threat of Fascism (57–8) – he was writing only days after the first Reichstag of the Third Reich had passed the Enabling Act for Hitler's dictatorship. Fascism, he notes, is a youth reaction, 'a declaration of the young right to be old', which offers a logical conclusion to the modern trend of timeless inertia by denying experiment of thought in order to supply mental and physical plenitude. The mirror image of this is Communism, in which a monolithic sense of progress creates a similar promise of plenitude: 'Fascism is the physical appeal; Communism is the mental appeal' (ibid). Yet the only equivalence between the two lies in the equivalence of physical and mental pulls Harrisson felt consciously. In fact, he is trying to generate a familiar trope of the period: a paradox or ambiguity similar to the Benjamin formulation concerning aesthetics and politics quoted in the Introduction to this book. This paradox arises because the Fascist-Communist dichotomy has been stated in the same terms as the Oxford-Cambridge dichotomy. By referring to the just-published anthology *New Country* including contributors from both Oxford and Cambridge – one of whom, as we shall see, was Madge – Harrisson offers a veiled hint at a resolution. To be sure, he is not exactly complimentary about the book: 'The revolutionaries are there but not the revolt. The book bears throughout the imprint of a cold hand, the atmosphere of stale tea' (60). However, at this point, Harrisson changes direction:

Yet this group of men from Ox and Cam (mentally led by W.H. Auden) is overwhelmingly in support of the future. As [Michael]

Roberts says in the preface: 'The present book, therefore, does not pretend to be "proletarian art"; it is important in so far as it is a picture of something that is happening: it shows how some of us are finding the way out of the individualist predicament ... It is to the younger men that we must look for an acceptance of a newer outlook.' [ellipsis in original] (ibid)

The attraction of this to Harrisson was the implied link between finding a way out of the individualist predicament and promoting mass acceptance of a newer outlook. As we can surmise from our speculations concerning his psychological outlook, the only way that he would be able to do the one, would be by simultaneously doing the other, but until he became fully self-conscious of that particular understanding both would appear impossible. It was precisely this self-realisation that gave him the confidence that he could change himself *and* his generation.

The last chapters of *Letter to Oxford* start to map out how such a change could take place. Harrisson begins this by shifting emphasis on to the 'Physical Oxen' and thus continuing the disruption of his initial categories. These are variously described as 'the Hollow Oxen, head-pieces stuffed with straw, Eliotwise', 'the British herd', 'the backbone of Empire' and toughs who 'will fight for Britain or anything else that they are told to. They will become Fascists incidentally' (64, 68, 71). This is very much an attack on public-school values in general from someone who in the past had been thrown into the Cam by Cambridge 'hearties' (Heimann 1998: 23) and whose red toe nails caused 'inconceivable' reactions in Oxford. Harrisson, of course, had never held these values and had always seen himself as an outsider whose ability to understand objectively how these systems worked enabled him to live freely within them. Therefore, he tempers his hate by analysis:

The Physical Ox talks very largely about extraordinary episodes connected with copulation: these are generally known as dirty jokes, and there can be no question that they are dirty. The psychology of smut is interesting. It is partly a frustration, a continuously offered insult to sex. More than that it is a way of belittling that factor, sex, which is inescapable even in the most civilised man or woman, and which offers a special impertinence to the athlete's controlled body. It is interesting that dirty jokes are an important feature of many savage communities and it is questionable whether the man who

cannot bear a little dirt is not the unhealthy specimen. In Britain
dirty jokes have in a queer way become localised and specialised in
certain strata of the community; they are entirely absent in many.
The presence or absence of dirt, and the type of dirt, provide master
keys to the understanding of these strata (69).

This is an explicitly anthropological analysis – including a footnoted
reference to Malinowski's *The Sexual Life of Savages* – which also functions as a deliberate put down. The sexual slight is reinforced by the
following paragraph, in which he argues that because the Physical Ox
nearly always starts with waitresses or whores there will inevitably be
inhibitions towards women of their own class but as 'the sort of
woman the Physical Ox marries is a backboness of Empire; so they are
the same; so it's comparatively alright' (69–70). This section amounts
to a wonderfully succinct functional social analysis of the colonial
public-school system which simultaneously calls for its dissolution:
'So I say – down with the Physical Ox. He is a decent animal: he should
not be allowed to remain one' (71). This early example of anthropology at home already shows clear deviations from the model of social
anthropology that was being established in Britain at that time by
Malinowski. Furthermore, Harrisson's manner of establishing his own
sexual credentials – 'My first heterosexual experience was not associated with a bed but with a clump of trees in a tropical jungle' (25) –
implicitly challenges the primary principles of that anthropological
model. A challenge echoed perhaps by the changing of the name on
the gate of the Gloucestershire cottage Harrisson and Bray rented to
'The Trobriands'. These challenges were to become more explicit later
in Harrisson's *Savage Civilisation* and the work of M-O, which in turn
triggered Malinowski's response in the long critical afterword to M-O's
First Year's Work.

Having written this epitaph for the ruling social system, Harrisson
paused to consider the emergent social class, 'the poor dumbox': 'It is
the grammar school; it is going to be a school teacher' (72). These were
the children of the new middle class, the sons and daughters of the
bank clerks and shop assistants: 'uncertain in their Wellso-Shavian
quietness'. Although Harrisson does not explicitly acknowledge the
affinity, they were people in an analogous position to the one he had
once occupied: simultaneously trying to understand and adapt to an
alien set of social values, while also reacting against them. They were,
as he suggests, an unknown quantity and the great question not just
for England but across Europe was: 'Will they be Fascists, Catholics,

Communists, Buchmanites, negatives, individualists, or just two-shirt men?' While M-O was never exclusively directed at this class – although the majority of its wider support came from it as we shall see – its attempt to offer everyone an anthropological understanding of the society they lived in was clearly of most immediate benefit to that section of society which had neither traditional ruling or working class values to fall back on – a condition which Kracauer had identified in Germany as 'homelessness'. A further motivation for M-O would be Harrisson and Madge's desire to keep this new middle class from becoming Fascist.

Harrisson's final burst is directed at his own class and generation as a whole, arguing that they do have the potential to change the world if they only broadened their horizons: 'Spend nights in the East End, be aware of homosexuals, stowaways, realise how little a man or woman can live on' (95). One of the aims of M-O would be to supply this information vicariously, but, paradoxically, even this process entailed an active participation in the passive experience of the mass. Harrisson already understood this in 1933, exhorting his fellows: 'Get occasional mob feeling, herd instinct, crowd sense. Universities don't give it to you except on Guy Fawkes day. Be mob-conscious, it is one of the grandest feelings. Realise, too, how far a mob can lead you to transcend, deceive yourself – war depends on mob-lunatik.' Finally he invites all the spiritually homeless to collectively imagine a new community using the everyday understanding that had developed from the need to cope with the social and psychological changes of the twentieth century:

> Life is exciting if only people will see it so. Not wonderful; just life ... It is easy to escape from the bondage of money, jobs, routine; you don't need a lot of brains, only guts and determination. You can see nearly everything in life without money, and it is the only way to see most of it whole. Keep on building up the structure from little things, from words and looks and lines and sudden incidents; not emotionally, but in a romantic-plus-rational way, neither one or the other' (96).

British Social Anthropology

Of all M-O's academic relationships, the one with anthropology has always been the most fraught, complex and ambiguous. It has never been as easy for anthropologists as for sociologists and literary critics to marginalise and criticise M-O because anthropologists such as Malinowski, David Pocock and, most recently, Brian Street have chosen to become involved with it. Then there are the awkward questions raised by Harrisson's own status as an anthropologist. Street, understandably anxious for anthropologists to take M-O seriously, has been cautiously critical in this respect distinguishing between Harrisson as someone who 'considered himself to be an anthropologist' and 'anthropological professionals at the time' such as Malinowski and Raymond Firth (Sheridan et al 2000: 79, 82). Conversely, Jonathan Benthall, director of the Royal Anthropological Institute, has emphasised the unoriginality of M-O but happily conceded that Harrisson was a pioneer of 'South-east Asian anthropology' (Benthall 2000). However, it is useful to remember that Harrisson was not the only link between M-O and anthropology. As we have seen, Madge read widely in anthropology while at school in the 1920s and it was he who liaised with Malinowski over the winter of 1937–8 and who attended his seminar at the LSE. Furthermore it should be remembered that the generation of the M-O founders – and prior generations, as acknowledged in Eliot's notes to 'The Waste Land' (Eliot 2002: 70) – were collectively influenced by Frazer's *The Golden Bough*. The particular effect of Frazer on the thinking about poetry of Jennings and Empson, which partly shaped M-O's first book *May the Twelfth*, will be considered in the next section of this chapter.

In the late nineteenth century, figures like Frazer and E.B. Tylor created 'a specialised discourse on social and cultural evolution' (Kuper 1996: x). In 1898, Alfred Haddon organised a Cambridge expedition to the Torres Straits, which included W.H.R. Rivers, and launched professional fieldwork in Britain. Rivers combined the evolutionism of Frazer and Tylor with German ideas of diffusionism. However, his views were in turn superseded by the tradition of functionalist anthropology established by Malinowski and A.R. Radcliffe-Brown, which was to become synonymous with British social anthropology from the 1920s to the early 1970s. Functionalism rejected diffusionism and evolutionism in favour of an intense synchronic study of a society and its culture based on an extended field placement by a trained worker who learnt the local language. Adam Kuper argues that it was the personal myth that developed around Malinowski which helped cement the approach into a code of professional practice. Having abandoned a career in science due to illness, a reading of Frazer brought Malinowski to the LSE to study anthropology. Brilliant progress took him on an anthropological mission to Australia, at which point the First World

War broke out. As an Austrian citizen, Malinowski was subject to internment but managed to spend the period in the Trobriand islands conducting intensive fieldwork by participant observation and living as one of the people. After the war, he returned to England and began the long struggle to establish anthropology on a modern footing. As Kuper concludes: 'the myth presents the classic story of a prophet. The false start, then the illness and conversion, followed by migration; the earth-shattering calamity – no less than a world war – leading to isolation in the wilderness; the return with a message; the battle of the disciples' (9–10). The truth is even more interesting: Malinowski began his first anthropological study while still at the University of Leipzig, where the ideas implicit to functionalism were present in the 'folk-psychology' of Wundt, under whom he was studying. He seems to have been planning fieldwork in Australia before the war and the Australians, rather than enforcing his stay, would have allowed his return to Europe but ended up funding part of his research. Furthermore, the research in the Trobriands was conducted over two year long spells with a sizeable gap in between and Malinowski was not successful in separating himself from European culture (10–13). None of which is to diminish Malinowski's immense achievements but merely to demonstrate that there was a gap between public image and actual practice. Malinowski obviously thought that maintaining this image was an important component of professional anthropology and hence his subsequent problems with Harrisson.

The idea of functionalism is that the relationship between different elements of a society is functional, that is to say that everything functions as part of a consistent whole. In accordance with this model, culture is seen as the way in which social needs are met. Therefore, the level of knowledge required to understand and explain how everything within a society is interconnected is that of an insider to the culture and hence the need for the anthropologist to live as one of the people being studied. It is no surprise that such a theory should develop in the long twentieth century from German roots, nor that Habermas would subsequently use functionalist terminology – in a predominantly structuralist account – to describe the development of the social sphere resulting from the structural transformation of the public sphere. The relation to *Verstehen* sociology is also clear in that the functionalist anthropologist is required to use a developed version of the everyday understanding of the natives of a society in order to interpret their social life. But, of course, the crucial difference is that these methods were not applied to British society but only to that of colonial subjects. This point was made forcibly in Perry Anderson's seminal essay of the 1960s, 'Components of the National Culture', in which he describes the success of British anthropology as the 'bizarre obverse' of the unique failure – in European terms – of modern British society to produce any form of classical sociology. According to Anderson, this was because the dominant class never faced a sufficiently serious challenge from within to make it worth theorising its own structure. However, because the same class was also ruling a third of the world, there was a displaced need for sociology: 'Colonial administration had an inherent need of cogent, objective information on the peoples over which it ruled' (265). Anthropology's utility for imperialism was aided by the privileging of image over practice in the Malinowskian model which served to obscure external influences. As Kuper observes, while Malinowski eventually came to see that the cultures he studied were not 'savage' but colonial cultures undergoing rapid change, the Trobriand monographs which made his name 'served as a potent example of how tribal cultures might be described as if they were "untouched"' (31–2). Combined with the characteristic synchronic approach, functionalism therefore gave rise to curiously static depictions of 'primitive' societies and, consequently, a later generation of British anthropologists would attack it not only for aiding the imperial project with information, but also for effectively denying historical agency to colonial subjects (for example, see Worsley 1970: 266–74).

Considering Harrisson's practice as an anthropologist within these contexts is a useful exercise for evaluating what is at stake in the debate over his precise anthropological status. On the Sarawak expedition of 1932, he 'learnt, for the first time, to appreciate an entirely different kind of people, the native pagan people of the Tinjar and Baram rivers. And they seemed to appreciate me, too. I found in these people something I had been looking for without success in the west' (Harrisson 1959: 154). Heimann shrewdly suggests that he was happy because his foreignness was accepted, allowing him to be himself, while his efforts to conform were met with unconditional praise. However, more importantly, this experience of simultaneously understanding and adapting to a different people fulfilled his search for a satisfactory combination of intellectual and practical approaches within an everyday activity – an everyday activity which was largely the preserve of anthropologists. Hence, while Harrisson's decision to stay behind in the New Hebrides after Baker and the official Oxford expedition party returned home in February 1934 was partly because it had become obvious that Zita was not going to leave her husband, it was mainly because it gave him the freedom to carry out the kind of anthropological work he wanted to. He was, in any case, conducting a population census designed to refute Rivers's thesis that the Melanesian population was dying out, but it was the attraction of meeting the cannibal Big Nambas of Malekula that particularly motivated him. His experiences resulted in the book *Savage Civilisation*.

Harrisson clearly draws on the Malinowskian model for presenting his results. The book begins with a strictly synchronic functionalist account of the village of Matanavat, apparently untouched by the Western world, told from the perspective of one of the villagers. However, he then proceeds to criticise the static view of Melanesian culture which this initial functionalism seems to support. Ironically, given that it is supposed to irrefutably mark him out as a positivist, it is his ornithological knowledge that Harrisson uses to demonstrate the holes in Rivers's thesis that the only changes in Melanesia were as a result of a single wave of immigration, by pointing out that 'the bird' whose common presence Rivers used to demonstrate proof of an immigrant common culture is, in fact, found in seventeen distinct forms, which could not possibly have evolved from one immigration. Therefore, Harrisson concludes: 'I believe, really, that every sort of impact and to and fro happened in the western Pacific, to build up this extraordinarily complex and locally differentiated culture' (338). Consequently he scorns the then widespread beliefs in 'racial decay' and Rivers's assertion that 'this dying out of the native races depends on the main on loss of interest in life'. Harrisson comments: '[Rivers] made this statement in 1921; it proved very acceptable ... The native, when confronted with white civilisation, lies down and dies' (269–70). This anti-colonial strain was backed up by colonial history and descriptions of the deprivations caused by the advent of the copra plantations. Yet it is the description of his own spell as a district officer under orders to 'attempt pacification by friendly means' (420) that does most to expose the complicity between functionalism and colonialism. Furthermore, Harrisson does serious damage to the professional image of the anthropologist in a number of carnivalesque episodes in the book. These culminate in his encounter with Douglas Fairbanks, Senior, resulting in a subsequent spell helping to make a documentary about the Malekulans. The attempt to film a man forcing his wife to suckle a piglet instead of a baby was only successfully accomplished by tying the piglet's snout and using a 'faked baby, made out of yam' (429–30). Coming at the end of the book, this can almost be read as a deliberate laying bare of the device by which the primitive is always somehow artificially constructed.

The problem with Harrisson's relationship with anthropology, then, is not so much that he falsely laid claim to its status, but that to a certain extent he held it up to ridicule – not least by taking the convention of writing as though from the point of view of the native through to its logical conclusion of voicing criticism of colonialism. Harrisson deliberately upset 'anthropological professionals' because they upset him, as can be judged by his later verdict on Malinowski: 'in those days we had all been stimulated by the anthropological researches of the chinless wonder pole, Professor Bronislaw Malinowski, whose methods have stamped themselves with such reduced but deadening effect upon the London School of Economics and much else in British social anthropology' (Harrisson 1959: 156). Yet, as this suggests, Harrisson had a slightly schizophrenic reaction to Malinowski, being equally inspired and repelled. As Kuper argues, for all the flaws that have emerged with Malinowski's work such as the static view of societies and the complicity with colonialism, there still remains a strong redeeming quality rooted in his focus on the strategies of individuals:

> ... his method carried a theoretical and even a moral charge. The ethnographer had to grasp the actor's point of view. If successful, the common humanity of Trobrianders and Europeans would be revealed. He passed on to his students an invaluable awareness of the tension which is always there between what people say and what they do, between individual interests and the social order, and directed their attention to the actor rather than the role, to the boasting, hypocritical, earthy, reasonable human being who could be found equally in Omarakana, Warsaw or London (Kuper 1996: 34).

In *Savage Civilisation*, Harrisson certainly grasped the viewpoint of the actor in New Hebridean society, describing kava drinking (275–7) and cannibal feasts: 'Every man must eat a portion. The taste is like that of tender pork, rather sweet. Some men are noted flesh-lovers and eat as much as a whole limb. The natives recognise, as I do myself, a peculiar greasy look about the eyes which characterises such men ... In general a small helping is enough...' (403–4). This account is described by Heimann as 'one of the most dishonest bits of prose Tom ever wrote for publication ... instead of a clear-cut statement, the text is ambiguous, confusion being caused by the way the narrative voice switches back and forth without warning from Tom's to that of a cannibal' (86). The point here is not the current orthodoxy that cannibalism has not existed in modern times, which Heimann rejects convincingly (80), but that it is highly unlikely that a foreigner would be allowed to attend such an event and, in any case, if Harrisson ever had participated in a cannibal feast, he would have declared so unambiguously. She accuses Harrisson of resorting to a literary technique, like that of Lowry, in the temptation to merge his voice with that of his informant and so give the impression that he was there at the cannibal ceremony. But is this that bad? One could argue that Harrisson's imaginative attempt to interact with cannibals is a more effective way of demystifying the rubric 'primitive' than a simple denial of the phenomenon existing. Indeed, by concentrating on the actor's point of view he demonstrates a common humanity rather than a representation of irreconcilable otherness. As he wrote later: 'I learnt there, to be a cannibal in feeling almost more than in fact. Or I thought I did; so that the difference did not matter (to me)' (Harrisson 1959: 156). Moreover, by achieving this through the use of a literary technique, he exposes the reliance of all anthropology upon the textual performativity of the act of writing: how else could Malinowski establish the actor's point of view? Harrisson's 'dishonest prose' in *Savage Civilisation* lies at the heart of a claim shared by himself and the British social anthropologists that: 'A person with a public-school background can understand the ideas and attitudes of a cannibal Malekulan as easily as he can understand those of a Welsh coal-miner or a Bessarabian' (342). Therefore, when Harrisson went one stage further in the merging of the voices of anthropologist and informant and claimed that they could speak as one in a vision of mass-observers speaking for themselves, of course Malinowski was appalled but also compelled to take a hand in M-O.

However, the British anthropological work that exerted the greatest influence over early M-O was Gregory Bateson's *Naven*, first published in 1936. The facts that the Cambridge educated Bateson's future was to be spent in the United States, where he married Margaret Mead, and that his approach was antithetical to the Malinowskian professional method, meant that he never exerted a great influence on British anthropology during the period dominated by the functionalist approach. Yet the latter of these two reasons made *Naven* attractive to M-O. Noting that anthropological orthodoxy suggests that a theoretical objective such as functional analysis should dictate the material investigated and collected by the fieldworker, Bateson insouciantly observes: 'But I had no such guiding interest when I was in the field' (Bateson 1958: 257). He goes on to state that his collection of facts was

carried out at random before he 'even dreamt of' his organising concepts – ethos, eidos and schismogenesis – but that, of course, he would
in the future use these theoretical concepts to organise his further
research. This was the model employed by M-O to justify its approach,
as for example in their August 1937 bulletin: '[observers'] work is going
to be the basis of a scientific theory of the areas, which we hope will
light up the whole nature of our society' (M-O A: FR A4). Furthermore,
this idea of social 'areas' was loosely based on Bateson's thinking, representing an intention by Madge to map out the eidos of society,
which would later re-emerge in his 1964 book *Society in the Mind:
Elements of Social Eidos*. Bateson defines eidos as 'a standardisation of
the cognitive aspects of the personality of the individuals' of a culture,
whereas he regards ethos as 'the culturally standardised system of
organisation of the instincts and emotions of individuals' (220).
Bateson argues that all stimulus-response type events have affective
and cognitive aspects. By classifying responses and how they are linked
together, it is possible to map out the affective aspects of a personality
and hence the ethos of the culture. By classifying stimuli according to
the responses they evoke it is possible to map out the eidos of a culture
literally, because what is being revealed is how the mental framework
of an individual in that culture classifies stimuli – i.e. if three different
stimuli provoke the same response, then it must be because they are
linked by the cognitive framework of the individual. As Bateson points
out: 'strictly speaking we arrive not at pictures of the individual but at
pictures of the events in which the individual is involved ... not to the
isolated individual but to the *individual in the world*' (274). In 'The
Normal Day-Survey' section of *May the Twelfth*, which is prefaced by a
quotation from Pavlov concerning stimuli and responses, Madge's
focus on finding a 'scientific classification' for 'social incidents' clearly
displays a determination to bring Batesonian anthropology home
(Jennings and Madge 1937b: 345–50, 370–1).

Popular Poetry and the 'Thirties'

Madge's poetry collection *Of Love, Time and Places* – incorporating his
two earlier collections *The Disappearing Castle* (1937) and *The Father
Found* (1941) and other poems written between 1932 and 1971 – was
published in 1994. Valentine Cunningham seized the opportunity of
reviewing the collection to brand Madge 'a second-hand merchant in
verse' and allowed only that the publication was 'an event of some liter-
ary-historical importance'. The implication was that this 'importance'

lay in the confirmation of the dominant literary reception of the 'Thirties', as prominently advanced, of course, by Cunningham himself: 'What this latest selection of his work by Madge himself illustrates is how a period of high cultural energy such as the 1930s can charge up the batteries of quite mediocre talents lucky enough to be in the right vicinity, and how such talents will be left poetically marooned once history and its major players have moved on' (Cunningham 1994: 4). What this review illustrates is an institutional process of cultural legitimation by which Madge, despite continuing to write poetry into the 1970s – and after (Calder 1996) – is used to demonstrate the apparently time-bound nature of 1930s preoccupations.

Generally, literary culture is divided between two spaces of radically unequal value. This is an idea that runs through the influential literary criticism of the twentieth century from Virginia Woolf's essays 'Modern Fiction' and 'Mr Bennett and Mrs Brown' up to Malcolm Bradbury's *The Modern British Novel* (2001). Bradbury, following Stephen Spender, describes the distinction as between 'the "Modern", experimental and avant-garde ... [and] the "Contemporary", which is fiction and literary art at its familiar work of exploring the world as in general we see it, and the way we live now' (xiii). Despite Bradbury's exemplary liberal claim to be fascinated by both traditions, he cannot escape the association of the 'Modern' with a transcendental literary value and the 'Contemporary' with a time-bound social relevance. The reason that the explicit debate in the 1930s between Modernism and Socialist Realism cannot be entirely reduced to the general opposition between the Modern and the Contemporary is that far too many writers bridge both categories. Literary criticism has only been able to evade this problem by focusing on select figures, notably Orwell and the Auden group, and regulating a whole cluster of people and movements to the category of social context, such as M-O, the English Surrealist Group, the Left Book Club, Penguin Specials, *Left Review* and proletarian writers. It is because Auden and Isherwood sailed to America in 1939, effectively renouncing politics or, at least, the politics of that 'low dishonest decade' (Auden 1986: 245), that they can be represented as the model for a literary construct known as the 'Thirties', which functions, despite appearances, to prove the general rule of the incompatibility of literature and politics.

This process is seen at its most subtle in Samuel Hynes's reading of Orwell's 'Inside the Whale' as creating what Hynes calls the 'Myth of the Thirties': 'On the whole, the literary history of the 'thirties seems to justify the opinion that a writer does well to keep out of politics' (cited in Hynes 1982: 387). Despite the fact that Orwell's comment is clearly ironic (see Hubble 2004: 36–7), Hynes treats it as a straightforward judgement. Therefore, by appearing to correct Orwell's invention of the 'Thirties', he actually uses Orwell to write the meaning of the 'Thirties' as *The Auden Generation* (see Hubble 2002: 15–6). Cunningham's *British Writers of the Thirties* (1988) performs a similar operation by misrepresenting and then expanding Orwell's account of the deleterious effects of exile on Henry Miller to encompass the whole decade: 'Orwell declared "... On the whole, in Miller's books you are reading about people leading the expatriate life, people drinking, talking, meditating, and fornicating, not about people working, marrying and bringing up children; a pity ..." There could scarcely be a sourer footnote to a period in which the hopes for what exile – especially "going over" to the working-classes – might bring to literature had run, so often, so deludingly high' (418).

Orwell, unlike the Auden group, represents a continuation of the literary politics of the 1930s forward into the postwar period where it links up with figures of the British New Left like Raymond Williams and E.P. Thompson. It is this reason combined with his enduring status that makes his reception so hard fought over and has led to so much, often conscious, misrepresentation. However, somebody like Madge, who also represents a continuation of the literary politics of the 1930s, is deemed sufficiently insignificant to be kept firmly in his time and place, as 'the ideal intellectual revolutionary simpleton of the ["Thirties"]' (Symons 1975: 88). This is despite the fact that Madge intervened during the 1970s and 1980s against the developing representation of the 'Thirties':

I was two thirds through Valentine Cunningham's thought provoking survey of fiction by James Hanley when I came up against a statement that 'he could occasionally sound just like Charles Madge praising the people's art that Mass-Observation techniques were supposed to uncover.' It is a shock to come unexpectedly upon one's own name like that, but in this instance the shock turned to incredulity because the Charles Madge referred to did not 'sound like' the person of that name with whom, for my sins, I have for the past sixty-six years been closely associated. True from 1937 to 1940 I wrote or co-authored various books, pamphlets and articles of Mass-Observation or based on M-O material. Rash claims were made in these, and unlikely projects mooted, but I doubt if there was in them much, if any, praise or dispraise of 'the people's art' and proletarian fiction ... Can Mr Cunningham give me a reference? Did he draw on some secondary source for his stereotype? Or did he simply invent it? If so, I would like to scotch it before it becomes an accepted part of 1930s mythology (Madge 1978c).

Madge was also unhappy with *Class, Culture and Social Change: A New View of the 1930s* edited by Frank Gloversmith when he reviewed it for *New Society* (Madge 1980). This ongoing re-evaluation of the 'Thirties' was generally dismissive of M-O and particularly hostile to its claim to be a form of mass poetry. Cunningham takes advantage of the unfortunate appearance of 'The Oxford Collective Poem' – a cut-up collaboration between twelve Oxford undergraduates, promoted by Madge – in *New Verse* to claim: 'These extremely trite eighteen lines were published by Madge ... to illustrate further the connection between Mass-Observation and poetry that he'd acclaimed in the magazine's previous number ... the Mass-Observation poem had about as much of the masses in it as ... a country house charade did' (Cunningham 1988: 339–40). Alternatively, Hynes attacks M-O for having too much of the masses in it: '... it is the *mass* in Mass-Observation that is numbing. The founders of M-O believed that the mass-consciousness could write a truer and better book than one man with his intuitions; *May the Twelfth* proves that they were wrong – writing will have to go on being an individual activity' (Hynes 1982: 286).

The dominance of this negative reception meant that, despite the genuine renewed interest in M-O from the mid 1980s onwards and the above-mentioned publication of his poems in 1994, the advent of the twenty-first century found Madge pictured above a caption of 'Where are they now?' in the *Guardian* newspaper's serialisation of Ian Hamilton's *Against Oblivion*. Hamilton's line of argument was that most twentieth-century poets are destined to fade into oblivion, if not already forgotten: 'Who speaks today of A.S.J. Tessimond, Charles Madge, Drummond Allison, and so on?' (Hamilton 2002). However, there are still people who speak of Madge and the existence of the M-O archive and the Madge papers together in the Special Collections of the University of Sussex Library ensures that there will remain people who speak of Madge for the foreseeable future. For instance, at the conference 'Mass-Observation as Poetry and Science: Charles Madge and his Contexts' held on 12 May 2000 at the University of Sussex, a number of papers specifically addressed Madge's poetry. Drew Milne argues that 'part of the interest of his poetry is that it is informed by a socialist poetics of perception that implies a revision of the literary historical terms conventionally applied to the poetry of the 1930s and poetry more generally' (Milne 2001: 66). This interest is compounded by his involvement with M-O because it both highlights the way in which poetry has become subordinated since the 1930s to increasingly domi-nant sociological approaches to culture, including cultural materialism, and raises the possibility of challenging that dominance. Milne notes how present-day Cultural Studies freely employs the term 'poetics' with respect to almost any cultural activity other than poetry – a trend which is accentuated in Everyday Life studies – and argues that this shows how the idea of poetry continues to exert influence beyond the apparent limitations of actual material poems: 'The work of Charles Madge provides an eloquent illustration of this paradox, to the extent that Madge's interest in poetry and poetics informs the project of M-O beyond the conventional terms of what is meant by poetry (68). The implication is that a revision of our ideas of poetics and our liter-ary historical sense of poetry would open up this paradox so that Madge's poetry and the mass poetry of M-O would both exert material influence in a less centralised culture. However, as Milne points out, the establishment of any such socialist poetics privileging 'collective inter-personal poetry' over hierarchical canons would substantially damage the work of individual poets: 'Without some shared sense of the value of poetic traditions, it is hard to see how poetry can persist as more than an idea' (70).

This problem is not entirely new. The changes in poetic perception which gathered pace during the early part of the long twentieth century led to movements such as Pound's Imagism – briefly discussed in the introduction – which seemed to reject significant elements of poetic tradition. These two apparently incompatible approaches were only reconciled by the formula, of impersonal identification with a timeless collective experience, described by T.S. Eliot in 'Tradition and the Individual Talent'. For Eliot, 'the new (the really new) work of art' is able to fit the tradition exactly because its introduction causes a simultaneous realignment of all the preceding works. At the same time, the original aspects of the poet's work are the least important: 'we shall often find that not only the best, but the most individual parts of his work may be those in which the dead poets, his ancestors, assert their immortality most vigorously' (Eliot 1932: 14). The enabling and empowering formula of this closed paradox has held poetry captive ever since, as can be seen from Milne's essay. The reason why Madge's career so effectively charts the limits of this paradox is not only his connection to M-O but also the connections between himself, M-O and Eliot. Eliot helped Madge following his departure from Cambridge by publishing some of his poems in *The Criterion* and, later, by arranging through a friend who was a leader writer on the *Daily Mirror* for Madge to have a job as a reporter with that newspaper. Madge's first two poetry collections were published by Eliot at Faber and, crucially, it was Eliot who managed to persuade his fellow directors that Faber should also publish *May the Twelfth* (see Madge 1987: 54, 60, 70). In return, it has been argued that Madge 'influenced the discourse of *Four Quartets*' (Calder 1996) and it seems more than likely that M-O influenced *Notes Towards the Definition of Culture*: 'Culture ... includes all the characteristic activities and interests of a people: Derby Day, Henley Regatta, Cowes, the twelfth of August, a cup final, the dog races, the pin table, the dart board, Wensleydale cheese, boiled cabbage cut into sections, beetroot in vinegar, nineteenth-century Gothic churches, and the music of Elgar' (Eliot 1948: 31). It is significant that through this book, Eliot has gained a central role in the history of British Cultural Studies, as seen for example in both a founding text (Williams 1990) and a key secondary account (Mulhern 2000), while Madge and M-O remain almost totally unacknowledged in this respect. This suggests that the problems Madge and M-O pose for poetry are connected to the problems they pose for Cultural Studies. It also suggests that there is a complicity between cultural materialism and Eliot's conception of modern poetry, such that rather than the former constantly threatening to subsume the latter, the latter depends on the former to maintain its distinctive singularity in the face of less exclusive conceptions of modern poetry. The best way of investigating this possibility is to follow the sequence by which Madge's interest in poetry and poetics progressed into M-O.

Starting with the influence of Richards, it can be seen that *Science and Poetry* reformulates Eliot's idea of impersonal identification with the tradition into a critical approach dependent upon 'both a passionate knowledge of poetry and a capacity for dispassionate psychological analysis' (9). In keeping with the latter, Richards carefully distinguishes between two kinds of poetic image: those of the intellectual stream 'which reflect or point to the things the thoughts are "of"' (13) and those of the active stream 'which deals with the things which thoughts reflect or point to' (14). At first sight, this distinction looks odd to present-day students and academics brought up on a diet of post-Sausurrean structuralist and poststructuralist theory based on the holy trinity of signifier, signified and sign because while the former reads like a description of a signifier and the latter of a signified, the implication here is that, rather than being part of the same process, they are parallel 'events'. Furthermore, Richards states that readers who only read or analyse at the intellectual level 'miss the real poem' and that this explains why so many people no longer read poetry (14). However, reading at the active level forecloses the possibility of ambiguity and misunderstanding arising from the intellectual level 'because the manner, the tone of voice the cadence and the rhythm play upon our interests and make *them* pick out from among an indefinite number of possibilities the precise particular thought which they need' (24). These elements of context and utterance were considered by Saussurre to be aspects of speech, *parole*, as opposed to the signifying system of language, *langue*. For Richards, the two need to combine in poetry so that rather than signifying the experience – 'the tide of impulses sweeping through the mind' – of the poet, the poem reproduces the experience in the reader by 'putting him for the while into a similar situation and leading to the same response' (26). Therefore, Richards argues that the words in poems are neither the cause nor the effect but a key to particular combinations of impulses.

It was on this subject that Madge first made contact with Richards in 1932, corresponding with him about what he called his *Ideolexicon*, which was a huge key to imagery in English poetry that Madge had drawn up as a guide to the hidden stores of knowledge in the poetic unconsciousness (Madge 1987: 33–4; Madge 1979: 11). Apparently Richards was not particularly impressed – although obviously impressed enough to take an active interest in Madge, persuading him to transfer to Moral Sciences and even helping him with his career even after he left Cambridge – perhaps he thought it too much of the past – too much a product of the magical world view which had been decaying for the past 350 years – for he had continued the argument in *Science and Poetry* by calling for a new order: 'a League of Nations for the moral ordering of the impulses' (35). This was his suggested response to the state of crisis he defined at the beginning of the book. The specific problem was that science had rendered many of the guiding principles that underpinned life in the past impossible to believe: 'For centuries they have been believed; now they are gone, irrecoverably; and the knowledge which has killed them is not of a kind upon which an equally fine organisation of the mind can be based' (60). As a solution, however, Richards argues that it is not necessary to believe in the guiding principles but merely to keep them for ordering our impulses as Eliot does in 'The Waste Land'. For the reason why 'poetry is capable of saving us', and only poetry, is because it retains the potential to give order to experience and to communicate that ordered experience to the reader, thus preserving the possibility of agency even in the state of 'mental chaos such as man has never experienced' which Richards predicts for the near future (55, 82–3). While this sounds straightforward enough, the real problem was not to fall into the temptation of conflating science with poetry because a poetry believed with the unqualified acceptance due to science would seem to transfigure the world: 'With the extension of science and the neutralisation of nature it has become difficult as well as dangerous. Yet it is still alluring; it has many analogies with drug-taking' (62).

Madge shows the influence of these ideas in his poem 'Instructions', written in 1932, which also reflects the impact of having read Auden for the first time. Like the *Experiment* group, he registers Richards's warning concerning premature transfiguration of the world:

> I am sorry
> That some workers in this field prematurely
> Published results and claimed exactness:
> We have underestimated the difficulty,
> And this is realized. (Madge 1994: 125)

While later in the poem, he acknowledges poetry's capacity to reproduce the poet's experience in the reader:

> This poem will be you if you will. So let it.
> I do not want you to stand still to get it. (127)

Madge sent this poem – or linked set of five poems to be precise – to Geoffrey Grigson and it was published in March 1933 in the second issue of Grigson's influential poetry journal *New Verse*, to which Madge was to become a regular contributor. However, in the meantime, Michael Roberts, editor of 1932's *New Signatures* and the forthcoming *New Country*, wrote to Richards on 16 December 1932, saying that Grigson had told him about Richards's new Cambridge communist poet: 'If it is not too much to ask, will you, at your discretion, invite the mysterious young man to send me something? We go to press on Jan 15th.'

Roberts's preface to *New Signatures* summarised the shared sense of situation at the beginning of the 1930s: the influence of Eliot and Richards and Auden's technical achievement of connecting contemporary industrial and urban imagery with subjective experience in everyday speech rhythms. He suggests that it may be 'possible to write "popular" poetry again: not by a deliberate patronising use of, say, music-hall material, but because the poet will find that he can best express his newly found attitude in terms of a symbolism which happens to be of exceptionally wide validity' (11). His examples of this possibility are not only Auden, Spender and Day-Lewis but also Empson, whose 'obscurity' is due to a 'necessary compression' of the logical analogy between ideas and the analogy between the corresponding emotional responses: 'In Mr Empson's poetry there is no scope for vagueness of interpretation, and its "difficulty" arises from this merit. Apart from their elegance, their purely poetic merit, they are important because they do something to remove the difficulties which have stood between the poet and the writing of popular poetry' (12).

Madge responded to Robert's invitation with the long poem 'Letter to the Intelligentsia' and an accompanying letter challenging Roberts's opinions as expressed in a recent review on Soviet Education:

> Can't we learn from our mistakes? Is there going to be the same pitiful waste of courage as before? I write without knowing if you have a program beyond 'We are to aim, not at the absorption of the upper classes (and the intellectuals) into the proletariat, but the absorption of the proletariat into a cultured middle class.' But if that is to be accomplished, we must have a schoolmaster's revolution ... For God's sake don't try to stabilise transition ...
>
> Why get wind-up as to the outcome of an English revolution? Of course it will be English: it will be the revolution of the English working class. You intellectuals find it hard to get rid of your conceit: 'fingertip of the consciousness of his period.' It's true – but one can't be the fingertip of two consciousnesses. Remember, though you have pushed the bourgeois consciousness neatly through the moves of the game, the proletarian consciousness is going to carry you off ... pointing finger and all – you may be a signpost, you'll have to be a surf-rider too.
>
> If you are all four [Roberts, Auden, Spender, Day-Lewis] what in my imagination you seem to be, this bolshevik criticism should do no harm. Am I on the same side as you? (Madge 1932).

The accompanying poem embodied these criticisms, rewriting the English tradition by rewriting Wordsworth – 'Lenin, would you were living at this hour: / England has need of you ...' – and rejecting the offer of what he takes to be only a modified public-school England:

> Yes, England, I was at school with you ...
>
> ....
>
> You are my one believed-in ghost. I've vaunted
> And venerated you, England, knowing you
> 'D rise from the dead and prove my superstition true.
>
> But we have left school now; we turn the pages
> Of a larger atlas ... (Madge 1994: 131–2)

There are other invocations of Lenin and calls for revolution amongst the published contributions to *New Country*, but it can be seen how Madge's arguments feed in to some of the positions taken by Roberts in the preface to *New Country*: '... it is for us to prepare the way for an English Lenin ... It is time that those who conserve something which is still valuable in England began to see that only a revolution can save their standards.' (M. Roberts 1933: 11).

Madge's position anticipates the hardline Marxist analysis of poetry that subsequently developed in the journal *Left Review* and is perhaps most fully realised in Christopher Caudwell's posthumously published *Illusion and Reality*. In this analysis it is the very illusoriness of poetry which creates the possibility of change in society but the problem is, as Caudwell details at some length, that bourgeois poetry expresses bourgeois illusions. While expressing the illusion of freedom undoubtedly helps to bring about its reality in bourgeois society, 'it is a freedom not of all society, but of the bourgeois class which appropriates the major part of society's products' (Caudwell 1977: 81). Therefore, Caudwell calls for a communist poetry which, instead of realising bourgeois freedom through expressing individual needs, will express a consciousness of social necessity and thus meet the challenge of 'refashioning the categories and technique of art so that it expresses the new world coming into being and is part of its realisation' (319). The problem is that nobody was producing such a communist poetry because proletarian values were only being accepted in every aspect of life except art, giving the poet a false distorting division: 'His proletarian living bursts into his art in the form of crude and grotesque scraps of Marxist phraseology and the mechanical application of the living proletarian theory – this is very clearly seen in the three English poets [i.e. Auden, Spender and Day-Lewis] most closely associated with the revolutionary movement' (314–5). These were the problems that Madge was aware of and struggling with. Commenting on Spender's essay 'Poetry and Revolution', he observes: 'The "danger" which Spender recognises is not the result of overdoing politics. Poetical poetry has suited in the past; political poetry very well. But poetical politics will never do ...' (Madge 1933b: 2–3). However, it is a mistake to link Madge too closely to Caudwell, especially with the fallacious argument that the more hardline communist the poet, the greater the slavish adherence to traditional metre (see e.g. Easthope 1979: 333). Madge's poetry displays a variety of forms, including prose poems and montages displaying the type of surrealist influences that Caudwell and the *Left Review* critics critically excoriated.

What gave Madge a different perspective from both his Communist peers and the Auden group was his connection to Jennings and Empson, both of whom he met through Raine. As we have seen, Roberts's preface to *New Signatures* ascribes to Empson a distinctive form of popular poetry. In keeping with the *Experiment* manifesto published in *Transition*, Empson was moving beyond Richards's position by not completely subordinating the intellectual stream of poetic images to the active stream: certain types of ambiguity created a correspondence between idea and emotion that simultaneously permitted precise interpretation and a transformative openness. Jackson describes Empson's *Seven Types of Ambiguity*, first published in 1930, as 'a kind of magical key to all mythologies' (Jackson 2004: 107), which has the effect of distinguishing it from the unmagical key that Richards was arguing for. The case for Empson arguing, contra Richards, that poetry can transfigure the world will be presented in the next chapter.

We know that Jennings shared and discussed ideas with Empson during this period. In particular, Jennings sent Empson a long letter concerning his theory of the Renaissance approval for Triumph:

> I think there is to be found in English poetry from 1580–1740 (from early Spenser to the revised 'Dunciad') symbolism or imagery (I take these terms as different but overlapping) derived from two things, (a) an extremely ancient philosophico-magical cosmological system, and from (b) the applications – partly ancient and partly contemporary (with the poets) – of this system, or branches of it. Note I take 'symbolism' to be derived from the abstract system, and 'imagery' to be derived from the applications (ritual etc) of it ... (cited in Jackson 2004: 387)

So far this is compatible with Richards's account of the magical world view: 'By the Magical View I mean, roughly, the belief in a world of Spirits and Powers which control events, and which can be evoked and, to some extent, controlled themselves by human practices [e.g. 'Ritual']' (Richards 1926: 47). Amongst these rituals, Jennings goes on to list 'the perpetual ritual of the lives of Renaissance princes (the progresses, "royal entries", masques etc)' (388). A list to which might be added coronations, suggesting part of the attraction that the coronation of George VI was to hold for Jennings. These rituals are imitations of the symbolic combats of the abstract system e.g. night and day etc:

> Between the microcosm and the macrocosm stands the King (see Frazer 'The Dying God'). As (a) the manifestation of the macrocosm & (b) the representative of the total tribe or community or of mankind itself, *his* combats are epic: definition of a hero. He is identified with the combats in both the sky and the earth. He is the sun & the wheat. His life is a perpetual ritual: both as microcosm & macrocosm he is *always* fighting combats ... (389)

As Jennings acknowledges, he is here drawing on Frazer's *The Golden Bough* – presumably also one of the sources for Richards's conception of the 'Magical View' – and one of its central tenets that the health and success of the king/god-man 'are a pledge and guarantee to their worshippers of the continuance and orderly succession of those physical phenomena upon which mankind depends for subsistence' (Frazer 1994: 262). Noting that it cannot be expected to be revived 'with the present state of agriculture', Jennings points out the abstract system is aristocratic and that, therefore, Eliot's referencing of Frazer in the notes to *The Waste Land* is in keeping with his royalism. So the question arises as to what value such systems and keys could possibly have for leftist intellectuals? On this point, Jennings echoes Richards again but with an important qualification: 'What is wanted *is* certainly a new system but it can't be found lying about' (390). He does not simply advocate retaining the system without belief for ordering our impulses because he argues that this is based on a misconception of the nature of the modern condition of chaos. Referring to an idea of Herbert Read's that there was a period before triumphs in which poetry and painting were protective rather than imitative and celebratory, Jennings argues that such conditions have returned with the difference that poetry and painting are no longer protecting us from nature (the macrocosm) but from ourselves (the microcosm). He later expressed this idea more explicitly in a radio broadcast:

> We've seen the way in which newspapers and short stories help us to deal with the *outside* world, but what about our lives by *ourselves*, Who is going to help us show off ourselves to ourselves? – because that is what we need. In fact what sort of language has man invented to deal with *himself*? Why, *poetry*, of course. When we repeat 'Tyger Tyger, burning bright', we're not talking about a *real* tiger, we're talking about ourselves, because with the poem we frighten ourselves – almost mesmerise ourselves – and at the same time we end up feeling as strong as a tiger (Jackson 1993: 260).

Apart from giving a clue as to what Jennings saw in M-O – i.e. the possibility of a kind of newspaper which would help us deal with ourselves – this amplifies his reference to Blake's 'I will not cease from mental strife' in connection with the need for a new system. The new system is clearly envisaged as an overcoming of oneself, as he instructs Empson: 'The difficulty of finding it, the battle against ourselves for it, the battle of the [*sic*] our four elements in our chaos, *are* the subjects you are looking for, as I see it' (cited in Jackson 2004: 390). This, in itself, raises questions as to what Empson's project is? Jackson dates this letter as 3 January 1930 based on Empson's recollection that he received it while at work on *Seven Types of Ambiguity* and probably used it although he could not say where (108). However, Jennings refers to Empson's work on 'Alice', which we know Empson to have begun after finishing *Seven Types* in the Autumn of 1930 (see Haffenden 2001: xxiii). Therefore, it seems at least plausible that the letter was written in January 1931 and the place where Empson 'used' it was *Some Versions of Pastoral* – perhaps in the chapter on 'Proletarian Literature' (23–4) and throughout the chapter on 'Double Plots' – the book in which the work on 'Alice' appeared and a major influence on M-O.

Alongside this idea of self-overcoming, Jennings had a particular conception of the image. Through his work as a theatre designer, a painter and, from 1934 onwards, a documentary film maker, he had developed an acute visual sense, especially with respect to different and multiple points of view. He understood that images were not concrete, not symbols, not pure forms, but the possibility of change, as he wrote during the early part of the Second World War: 'an object cannot immediately exchange its being with another object. But an *image* of an object is immediately exchangeable with another *image*. An image of a horse can become an image of a locomotive' (Jennings 1982: 46). Years later, after Jennings's death, Madge discussed this idea of the image and pointed out that in terms of technique, it involves metamorphosis rather than literary substitution. This is to say that rather than freely signifying and so perhaps evading the imposition of external meaning, the image holds both one meaning and the sense of its self-transition to another meaning, equally its own. Accordingly, in either poetry or painting: 'Paradoxically solid and fluid, the images are moments in the flow of human experience. The shape is solid, but the line that encloses it is fluid, as it awaits the next metamorphosis.' Madge went on to conclude: '[Jennings's] aim was to seize and create "mutations in the subject", liberating human perceptions from the literature that surrounds them' (Madge 1982: 48). This simultaneous consciousness of chaos and order, flux and stasis, avoids the twin pitfalls of freezing meaning to an absolute or evaporating it into an endless play of differences.

In the Autumn of 1935, Madge and Raine moved to Blackheath to be near Humphrey and Cicely Jennings, who lived there because Jennings was working at the nearby studios of the GPO Film Unit. Immediately, Jennings and Madge began to collaborate on a montage made from accounts of the Italian invasion of Abyssinia, the colonisation of North America and various newspaper reports, published in the December 1935 issue of *Life and Letters* under the title of 'The Space of Former Heaven'. The finished text combines Jennings's imagist technique and Madge's political purpose as ideas of a new symbolic order based on self-overcoming merge with a communist poetry expressing the new world coming into being. The net effect is to show the continual combat between such apparent opposites as primitive tribesmen and European civilisation sliding into an unstable metamorphosis between the two conditions, while the mock pompousness suggests a document from an alternate history governed by a more democratic abstract symbolic order.

It is not simply the case that Jennings influenced Madge. A cosmological abstract order features in many of Madge's earlier poems such as 'Man under Taurus', 'Solar Creation' and, most successfully, 'The Hours of the Planets' – a playful 'Waste Land' – which culminates in a section entitled 'The Key'. 'The Space of Former Heaven' – the humour of which seems more characteristic of Madge than Jennings – merely registers a gradual development, already discernible in his work, towards montage prose poems focusing on spaces, or more precisely 'Landscapes', as a sequence is named. These pieces have the compressed feel of three dimensions squashed into two. The resultant ambiguity of images simultaneously acting on more than one axis generates the potential for creative metamorphosis, as can be seen in 'Bourgeois News':

> Floods are frequent because the rivers of Britain have been neglected for a century. Positive movements of transgression carry the sea and its deposits over the lands, drowning them and their features under tens or hundreds of fathoms of water. Efforts to advance the prosperity of the country should be directed towards building on the foundations already laid by the native himself, rather than to hazardous introductions or innovations. Commercial possibilities are not clearly and courageously visualized, and the new ventures are often the product and concern of individuals facing the traditional difficulties of lonely pioneers. The indoor staff remains comparatively small. The vigour of mountain building, of volcanoes, and of other manifestations of unrest, has shown no sign of senility or lack of energy. An operator received concussion and a wound on the head from a cast-iron cover blown off a 60A switch-fuse box (Madge 1936b: 7).

As Rod Mengham observes: '["Bourgeois News"] is in effect a report on the British, who have been made the object of the kind of scrutiny which an anthropologist would bring to bear on the so-called primitive rituals of an exotic tribal culture. In this it pre-dates but anticipates how anthropology will "begin at home"' (Mengham 2001: 32). The significant word here is 'British'. Madge, realising that 'England is fallen', spent 1936 mapping what might be called 'The Space of Former England' which prefigures, but has not yet become, a British landscape. Mengham, employing the analogy of the backing music that Jennings used to overlap the cutting of scenes together in his wartime films, suggests that the movement of the prose in the montage-style reports of Jennings and Madge provides a 'soundtrack' binding together the otherwise 'disparate elements of the text' (33). This raises the possibility of the 'soundtrack' of 'Bourgeois News' becoming an authoritative voice-over precluding the reader from making alternative connections. Mengham argues that this technique has the potential to become fully authoritarian as, for instance, a similar type of political oratory does in parts of Auden and Upward and, it is implied, M-O itself, with its 'method of speaking *for* subordinates, rather than letting equals speak for themselves'. Finally, he concludes 'in "Bourgeois News" it may be that autocratic and democratic perspectives are still being held in tension. "Bourgeois News" is one of those texts in which history is Janus-faced; a movement which could go either in this direction or in that' (33).

This poised ambiguity can be understood in terms derived from Richards: the authoritarian 'soundtrack' is a form of the intellectual stream of images, which is contested by the symbolic oppositions of the active stream of images such as those between flood deposition and volcanic eruption, to pick examples from the extract quoted above. As in 'The Space of Former Heaven', the parody of official forms, such as newspapers, textbooks and government reports, helps lay bare the device by which these forms are used to hide symbolic and social oppositions:

> In the course of yesterday, the successive bulletins were of a more reassuring nature: 'The most probable cause is the present state of flux in native life, the disappearance of tribal discipline, and the results of undigested education. It is possible to change some factors; it is not possible to change geography.' These reassurances did not everywhere produce the desired effect (Madge 1936b: 9).

The result is more than a simple satire, because the subtle shifts in tense and tone hint at exactly such hitherto unimagined transfigurations of social space:

> The ropes snapped, and the flying-boat drifted for 100 ft., and sank in 60 ft. of water, where it remained. As the day wore on, and the anticyclone began to withdraw to the Continent, three quarters of those present made for the door. There was no panic. They could go on their way peacefully, because they were strong. (10).

Jennings employs similar devices in the 'reports' he wrote over the second half of the 1930s and in his writing and editing for the London scenes in M-O's *May the Twelfth*. However, there is one important difference between Jennings's work and 'Bourgeois News' and that lies in the way that Jennings resists closure either by producing only fragments, as with the 'reports', or by employing an implicitly cyclic diurnal structure as in *May the Twelfth* and some of his wartime films (see Winston 1995: 106). By contrast, the form of both 'The Space of Former Heaven' and 'Bourgeois News' – as opposed to Madge's more fragmentary 'Landscapes' – is much closer to that of a short story. It may well be this combination of narrative form and prose movement that creates the elements of the authoritarian 'soundtrack' that Mengham detects. If so, the significance of 'Bourgeois News' might lie not so much in what it tells us about spring 1936, as in the way its form prefigures the wartime crystallisation of the poised 1930s tension between authoritarian and democratic perspectives into the fixed structures of the postwar British State.

The next logical move for Madge was to take the final step from the last vestiges of individualism in his prose poems and progress to the 'popular poetry' anticipated by Roberts in *New Signatures*. A series of notes under exactly this title, 'Popular Poetry', in one of Madge's notebooks, calls for 'Coincidence Clubs: groups in colleges, factories, localities' to study the press and advertisements, and be involved in 'exercises for imagination'. Under a list of 'Plans for PP' come 'First Text Book of PP', lectures, training courses, 'mimeographed record sheets', delegate conferences and 'PP Newspaper on mass basis'. Possible slogans include 'Newspaper Active' and 'Mass Science'. A list of potential contacts includes Empson, Richards, Edgell Rickword, Gavin Ewart, Roland Penrose, John Betjeman and Jennings – suggesting that the notes were compiled by Madge on his own. The concepts 'Materialism' and 'Class Consciousness' indicate the Marxist objectives of the putative organisation, given a characteristic spin: 'English people must learn *to like* their surroundings before they can change them. The curse of misery is apathy.' A final note promises: 'Movt can cut the Gordian Knot' (Madge 1936c).

This has to be seen as an early – probably the earliest – plan for what was to become M-O. The notes are significant for making clear certain influences, for showing that the plans for M-O preceded the Abdication crisis – which would surely otherwise be mentioned – and establishing that the original conception was of a political movement. Unfortunately the notes are undated, the last entry in a notebook started on 9 December 1934, but the fact that Penrose's name is placed next to the initials S.G.G.B., means that they must have been written after the formation of the Surrealist Group on 7 July 1936 – these initials seem to be Madge's own designation for what members of the movement referred to in print as 'The Surrealist Group in England' (Surrealist Group in England 1936) or the 'English surrealist group' (Roughton 1936: 74). Given that the Abdication crisis became public at the end of November, Madge must have written the PP notes over the summer or autumn of 1936. It is perhaps indicative that Madge's essay 'Press, Radio and Social Consciousness' uses the term 'popular poetry' to describe the story of the 'Human Mole' printed in the *Daily Mirror* of 26 October 1936 (Madge 1937e: 161). In any case, he had begun the discussions that led to the foundation of M-O by 1 December because David Gascoyne later recalled that returning by train – from Blackheath to Waterloo, en route for Richmond – after one such meeting: 'I saw a great glow in the sky, which the next morning I found to have been the great fire of the old Crystal Palace, a "happening" which greatly struck the popular imagination at the time ...' (Gascoyne 1980: 9). Gascoyne was a member of the Surrealist group, while Stuart Legg, another participant in the discussions, was a member of the documentary movement; Jennings was a member of both. The interplay between intellectual and active streams of images in both of these movements, with their variable mixes of democratic and autocratic tendencies, created the final, most immediate, context for M-O.

3
The Intellectuals and the Masses

Profane Illumination and New Objectivity: English Surrealism and British Documentary

John Carey's populist broadside, *The Intellectuals and the Masses*, related M-O to his central thesis in no uncertain terms:

> Eliminating the humanity of the masses can also be effected by converting them into scientific specimens. This was the enterprise undertaken in the 1930s by Mass-Observation. Tom Harrisson ... and his team of middle-class observers based themselves in Bolton ('Worktown' in Mass-Observation code) and mingled with the natives, collecting data on local customs ... Observers were instructed to use an impersonal notation when identifying human specimens. The formula 'M 45 D', for example, meant a man of about forty-five who looked or sounded unskilled working class (Category D).
>
> Amateurish and innocent as Mass-Observation now seems, its employment of a scientific model for the purpose of segregating and degrading the mass had a sinister counterpart in the assimilation of the masses to bacteria and bacilli (25).

In an endnote referring to this passage, Carey continues 'In so far as Mass-Observation proved anything about the mass, it was that the mass did not exist. Its volunteers on the 'National Panel' seem to be just as individual as anyone else ...' (221). This rather lame conclusion sums up the major shortcoming of the book: the insistence on seeing the idea of the 'mass' as purely an invention by intellectuals (see 210). The problem is that this formulation leaves nothing for Carey to argue for, other than certain individuals: 'Arnold Bennett is the hero of this book' (152). As such, Carey's position is less likely to gather adherents than the one he attacks which, in its purest form as advanced by Leavis – the idea that mass civilisation is the enemy of minority culture – offers all the temptations of an oppositional community united by common values and shared moral purpose.

Leavis contested the processes of modernity during which the traditional organic community had been devastated by the logic of economic change and the consequent social transformations. He saw no worth in the prospect of founding an alternative society by accelerating and extending those transformations, dismissively categorising 'the Marxian future' as 'vacuous, Wellsian and bourgeois' (cited in Mulhern 2000: 15–7). The fact that Carey's split account of H.G. Wells as being both for and against the masses does no more than offer the suggestion that the relationship between the individual and society might have hidden complexities, illustrates how far his overall position – despite a superficial reversal – remains stuck in a Leavisite paradigm; making no argument for any positive conception of the 'masses'. Instead of being portrayed as a schizophrenic, Wells might be seen as the representative of a wider mass cultural politics in which autocratic and democratic perspectives were held in tension. Wells wrote for the clerks and shopkeepers – as Carey acknowledges: 'the Mr Pollys and Hoopdrivers and Kippses' (145) – and was linked with the political journals targeted at this audience such as Orage's *New Age* and the *New Statesman*. These journals established a very different paradigm than Leavis for the interaction of intellectuals and the masses; one that 'chimed with the aspirations of thousands of individuals and small groups throughout the country who were uncommitted, progressive and for the most part young' (Carswell 1978: 35). Although by 1937, when M-O was launched through its pages, the readership of the *New Statesman* represented a more homogenous professional middle class than it had done before the First World War, the social profile of the Mass-Observers recapitulated that original Wellsian constellation, because they were also being recruited from appeals and articles in other journals and newspapers such as the *News Chronicle* and the *Daily Express* – via Harrisson's contacts (Heimann 1998: 133). This is where it is important to distinguish – unlike Carey – between the passive subjects of M-O's participant observation in Worktown and the self-selecting volunteers who became day-diarists in 1937 and, later, members of the 'National Panel'. Tom Jeffery has classified these volunteers as preponderantly lower middle class like 'the Mr Pollys and Hoopdrivers and Kippses':

> Few were in full-time education much past the age of sixteen, although many had won scholarships from elementary to secondary school. Those who did go to university almost without exception returned to the lower middle class world as schoolteachers ... The unmarried diarists tended to live with their parents through their twenties ... The homes of the lower middle class diarists tended to be in the suburbs of large cities, often in streets which were just beginning to go noticeably down in the world, streets which would have been eminently respectable when the twenty-five year old diarist was born. The diaries contain accounts of the work of clerks, shop-assistants, draughtsmen, many school teachers, commercial travellers, and laboratory technicians ... Few started before half past nine, most took a mid-morning coffee break outside the office or shop, most took at least an hour for lunch and many were off home by four thirty; few stayed at work after five (Jeffery 1978: 29).

In later work, Jeffery has refined this classification, noting that observers predominantly chose to define themselves as belonging to a 'new lower middle class' defined largely by work, incomes, suburban homes and education, often technological. He argues that the involvement of this class in M-O as with their reading of Penguin books – launched in 1935 – and the *News Chronicle* and their frequent membership of the Left Book Club – founded in 1936 – represents the re-emergence following the 1935 General Election of lower-middle-class progressivism and radicalism that had survived from the prewar period only to be submerged by the great depression and subsequent unchallenged rule of the National Government. While the majority of the lower middle classes remained passively loyal to conservatism, this 'new' fraction, often rooted in occupations such as civil servants, teachers and railway clerks that had unionised and established secure national bargaining agreements by the early 1920s, were also re-politicised by events abroad in Germany, Spain and the Soviet Union (Jeffery 1990: 70–1, 75–6, 85).

It can be seen how this history loosely mirrors that of the new middle class in Germany as outlined variously by Lederer, Marschak and Speier. However, because long-twentieth century structural transformation was most concentrated and advanced in Germany, where it led to the widespread experience of a depoliticised everyday life, the particular problems of repoliticising everyday life are shown by German history in their most extreme forms. For example, Rudy Koshar argues that the waves of politicisation that caused the rise and fall of the Weimar republic 'would have been less rapid and dramatic had large parts of the German population not already received an indirect political education in the everyday life of voluntary groups' (Koshar 1990: 36). He also notes that 'the evidence suggests that the fractiousness of German voluntary groups resulted not from petit-bourgeois selfishness, but from the penetration of elite models and practices throughout everyday communications networks (35).' This was a consequence of that double movement described by Habermas in which the public and the private merged in the simultaneous deprivatisation of property holders and pseudo-privatisation of workers. The resultant 'refeudalisation' was more than just the rejoining of state and civil society, it also entailed a literal reintroduction of feudal values from the Prussian bureaucracy into what had been a relatively emancipated nineteenth-century public sphere. Such practices could do nothing but cause fractiousness in a social sphere dependent on a functional equivalence between the different but necessary forms of participatory practice which constituted it. Therefore, the implicit possibility of social transformation was lost and social consciousness divided between its nineteenth-century bourgeois and proletarian poles in the attempt to escape the refeudalised social sphere, leaving only the salaried masses stranded in the middle as a trace of the effacement of the future.

These are the reasons why Kracauer remained sympathetic to the salaried masses and both attacked new management for adopting outmoded values and arrived, as Benjamin *approvingly* noted in a contemporary review, 'at a critique of trade unionism' (Benjamin 1998: 110). This approval was awarded because Kracauer's awareness that the unions were simply another part of the total social system that structured everydayness, accorded with Benjamin's rejection of the idea of the industrial proletariat as the benchmark of authentic existence. While the proletariat might be the living embodiment of the irreducible everyday, the reality of this was, as Marx had once written, 'that the worker feels that he is acting freely only in his animal functions – eating, drinking and procreating, or at most in his dwelling or adornment – while in his human functions he is nothing more than an animal' (Marx 1992: 327). The existence of this irreducible everyday in contained spheres provided no more than an escape from the otherwise enforced necessity of existing in the face of the contradictions of everyday life. Therefore, the attraction of the new middle class to Benjamin, from his position of regarding 'the production of a proper consciousness ... [as] the primary task of Marxism', was that they no longer had this retreat open to them precisely because they had been completely absorbed into the functional performance of the social sphere: 'Their more indirect relation to the production process finds its counterpart in a far more direct involvement in the very forms of inter-personal relation which find their counterpart in this production process' (110).

Benjamin's argument was directed against those who insisted on absolute identification with the proletariat: 'This left-radical school [*die neue Sachlichkeit* – the New Objectivity] may conduct itself how it likes, it can never eliminate the fact that even proletarianisation of the intellectual almost never creates a proletarian' (113). Against this, he held up the model of the Surrealists: 'In reality, it is far less a matter of making the artist of bourgeois origin into a master of 'proletarian art' than of deploying him, even at the expense of his artistic activity, at important points in [the] image space. Indeed, mightn't the interruption of his 'artistic career' perhaps be an essential part of his new function?' (Benjamin 1999: 217). Similar statements would later figure in the early writings of M-O: '[The observers] produce a poetry which is not, as at present, restricted to a handful of esoteric performers. The immediate effect of Mass-Observation is to devalue considerably the status of the "poet". It makes the term "poet" apply not to his perfor-mance, but to his profession, like "footballer"' (Jennings and Madge 1937a: 2). The roots of this understanding lay in Jennings and Madge's close knowledge of the debates and differences within and between the English adherents of the two continental movements.

The primary importers of the ideas of the New Objectivity into Britain were John Grierson, the leader of the British documentary movement, and Auden and Isherwood (Willett 1978: 224). As John Roberts demonstrates, all these movements were predicated on the impact of the Russian Revolution, which offered a model for politicising and directing the ongoing transformation of the social sphere: 'Building a new world out of the ruins of absolutism and industrial backwardness, the working class was in a position to redefine the categories and boundaries of the real. The "everyday" then became open to an extensive and unprecedented politicisa-tion, insofar as its intellectual redefinition became co-extensive with social transformation itself' (Roberts 1998: 16). Consequently the old Marxist problem of false consciousness appeared to have been resolved because proletarian everyday consciousness was now identical with social consciousness as a whole. Furthermore, Soviet photographic and film montage reworked naturalistic categories of time and space in order to present this proletarian consciousness as unified across temporal boundaries from the past into a future that had already begun. It was exactly this demonstration of the unity of everyday consciousness that Benjamin criticised as premature because it marked a retreat from the political contestation of everyday life, rather than a genuine effort to resolve social contradictions (see Roberts 1998: 30–2). Nevertheless, representations of the everyday as proletarian necessarily became a form of social contestation within the German contexts of Weimar social injustice and the rise of fascism and, as Roberts notes, 'a cult of the "factual" grew up around photographic practices' (47). Implicit in the development of the New Objectivity, therefore, was a shift in political emphasis from social transformation to defensive anti-fascism. Transition to Britain marked a final separation from the original revolutionary politics of social transformation for practices that were to become associated with the documentary movement in particular and Popular Front approaches in general.

Roberts suggests that Grierson's characteristic approach was 'to tell stories of the "everyday"' which portrayed ordinary events as poetic acts and thus served the need to culturally legitimate documentary films as distinct from commercial cinema and the taint of mass civilisation (59, 62). Yet, the accompanying shift in class content 'from working-class interests to the general human good', while admittedly opening the way to the hegemonic postwar form of documentary as a 'liberal project of education from above' (60), also potentially opened the way to representations of a mass society in which the categories of everyday life were contested. However, in practice this development was always restricted by the need for the documentary movement to comply with the wishes of their state sponsors. Grierson's film unit was first set up in 1930 as part of the EMB following the amazing success of *Drifters*, a film Grierson made for them about the Scottish herring fleet. Interestingly, two of the civil servants directly involved with the use of film by the EMB were subsequently connected with M-O. Stephen Tallents, the EMB secretary who had formerly served as the secretary to the Cabinet Committee which dealt with the General Strike, and Frank Pick, responsible for publicity after having been extremely influential in that capacity at the London Passenger Transport Board, were both later to hold the post of Director General of the MOI. While at the

EMB, Tallents and Pick developed an innovative public relations policy designed, in Pick's words to 'establish a background and to develop gradually a permanent Empire consciousness rather than to try and create an immediate demand for goods' (cited in Swann 1989: 24–5). This model, with its scrupulously kept distance from advertising, was to be perfect for the detached politics of practice which formed the ideological core of the British documentary movement and even allowed films to be made directly for commercial concerns.

In September 1933, the entire unit transferred with Tallents to the public relations department of the GPO and shortly afterwards took over a small studio in Blackheath. Jennings, in need of employment to support a wife and child, joined the GPO Film Unit in 1934, probably through his Cambridge friend Stuart Legg, who was already working as a director for Grierson (Jackson 2004: 144–7). Fellow unit members and subsequent leading figures in postwar documentary, Arthur Elton and Basil Wright were also recent Cambridge graduates. As with his earlier theatrical career, Jennings designed sets and even acted, but by the end of the year he had edited and directed his first films. He worked for the unit off and on until the end of the decade and then throughout the war under its new guise as the Crown Film Unit. However, he never got on with Grierson and the departures from the prescriptive norms of documentary which characterise his later famous films led to a controversial future reception history among his fellow documentarists. These disputes highlight the limits and paradoxes of the British documentary form by focusing on its distinctive poetic strand. While Alberto Cavalcanti judged Jennings a 'film poet', Grierson would only concede that he was 'a minor poet' and chose to consider the 1936 collaboration with Auden, *Night Mail*, as the end of the poetic documentary approach: 'We worked together and produced a kind of film that gave great promise of very high development of the poetic documentary. But ... we got on to the social problems of the world, and we ourselves deviated from the poetic line.' On the other hand, Legg suggested 'that Humphrey probably represented the final development of the poetic line, as we saw it in this country' (all cited in Sussex 1975: 110–1, 79).

A consideration of Jennings's 1939 film *Spare Time* illustrates why his work was to provoke such fundamental disagreement. Although *Spare Time* was made after Jennings had left M-O, he stayed with the Worktown unit – by then run by Madge (Madge 1987: 98) – while shooting scenes in Bolton and, therefore, it is the nearest there is to an actual M-O film. Indeed, the shot of a young woman making herself up in the mirror is of one of the local mass-observers (see G. Thomas 1939). The film was intended as the second part of a joint submission, *British Workers*, to the New York World's Fair (Jackson 2004: 210–11). As such it focused on recreations in three areas dominated by different industries: Bolton and the North West (cotton), Sheffield (steel) and South Wales (coal). One of the Northwest scenes, that of a children's kazoo band, has become notorious. Years afterwards, Wright recalled the reaction of others in the film unit: 'the general feeling when the film first came out ... was that Humphrey seemed to show, in our opinion, a patronising, sometimes almost sneering attitude towards the efforts of the lower-income groups to entertain themselves' (cited in Sussex 1975: 110). He went on to note, though, that he had subsequently revised his opinion, particularly in the light of Jennings's later career and his postwar critical standing: 'I think we were a bit too doctrinaire in our attitude in those days and, looking at the film again today and remembering that it was made for Mass-Observation, which was a particular organisation with a particular purpose, I think that perhaps we may have missed the point.' However, at least one recent critical account has endorsed those early reactions: '*Spare Time* ... contains what are easily the movement's most alienated and alienating images of the working class in the prewar period' (Winston 1995: 53). Of course, it is precisely the use of images which is under dispute in the absence, as Jackson patiently points out, of distorting angles and sarcastic cut-aways: 'it is this very policy of abstention, one suspects, that makes viewers uncomfortable, since it presents them with a smooth surface in which they are able to see, not Jennings's preconceptions, but a reflection of their own' (Jackson 2004: 215).

Unlike most of his colleagues, Jennings refused to follow the instructions of his erstwhile mentor Richards concerning the necessity of subordinating the intellectual stream of poetic images to the active stream. As a consequence he could never achieve the distinctive closed system of representation by which documentary habitually portrayed the working class as either everyday heroes or unjust victims. His films with their metamorphic images are imbued with an ambiguity that superficially leaves him open to the accusation of subordinating collective needs to a self-serving cold intellectualism. However, what they did was invite audiences to identify with a sense of the possibilities of social transformation rather than with the reassuring comfort of an everyday plenitude. Hence, in *Spare Time*, we experience a series of images which simultaneously represent mass and individual identity: as a football crowd leaps to cheer a goal, a pair of hands reaches down over the hoarding to throw the ball back; as couples swirl around a ballroom to dreamy swing, one young woman reaches down to adjust her shoe strap while continuing to dance in time; as the members of a male choir take off their coats and enter a room for rehearsal, a pianist plays without missing a beat even as she is helped out of her own coat. Collective activity and individual agency are combined in a free identity, which is only surrendered at the call of the works' siren.

Part of the problem for the contemporary reception of Jennings's work was that a critical discourse capable of exposition was only just beginning to appear in relation to the separate movement of surrealism. One of the earliest articles in English favourable to surrealism was Madge's acute review of the *Petite Anthologie Poetique du Surrealisme* in *New Verse*. He suggests that the value of the experiments in automatic writing conducted by Breton and Aragon resides in their demonstration of Lautréamont's argument of the necessity for simultaneous multiple personality: 'The Ferdinand-Ariel relation of what one may call the Poet and his Muse is only attained at certain lucky moments, as it were anonymously and without authorisation. At such moments the spell is complete and liberation is possible' (Madge 1934: 14). It is difficult to think of a more fitting description for Jennings's achievement in *Spare Time*. Madge continues by noting that: 'In comparing one surrealist with another, the only possible index is that of technical skill in performing these manoeuvres of dissociation.' The statement could also be employed profitably by substituting 'documentary filmmaker' for 'surrealist'.

Madge was unusual in not sharing the heavily critical line on surrealism later to be taken by his fellow contributors (Blunt 1936; Jackson 1936; Lloyd 1937) to *Left Review*, the semi-autonomous house journal of the communist literary intelligentsia founded in 1934. This was, in fact, quite a cosy rivalry centred on the shared environs of the Arts Café at 1 Parton Street, the bookshop opposite and the office of *Left Review* at 2 Parton Street. However, it was the more independent Caudwell who expressed the objection to surrealism most succinctly by characterising it as – curiously echoing Benjamin – 'the final bourgeois position' (Caudwell 1977: 280). According to him, while sincere modern bourgeois art, such as that of Lawrence and Joyce, breaks apart under the tension caused by the increasing gap between bourgeois consciousness and the exploited class, into a maelstrom of fragmented consciousness; all surrealism attempts is to haul itself up by its own bootlaces through positivistically recording and then embracing those fragments as the key to the unconscious and the possibility of a dream state where 'men float into air, cut loose from both subject and object' (265). Of course, this very exploitation of the image space created by the structural transformation and separation of bourgeois and proletarian consciousness was precisely the attraction of surrealism to thinkers like Benjamin, Jennings and Madge. The particular genius of the latter two was to link surrealist technique to the task of helping the stranded new middle class find the appropriate social consciousness for themselves. However, it was this aim which also distanced them from surrealism in England and led, in particular, to the split in the English Surrealist Group.

While the Group was only formed in 1936, following the success of the International Surrealist Exhibition in London that summer, one line of its domestic agenda had been set out in advance by Herbert Read in an article entitled 'Why the English Have no Taste'. Apparently, the English, conditioned by capitalism and puritanism, are obsessed with a normality defined by common sense and the English sense of humour: 'mental laughter, caused by an unconscious disturbance of suppressed instincts' (cited in Remy 2000: 16–17). English Surrealism was to act as a corrective to the lack of freedom resulting from this conformism to normality. An approach exemplified, according to Michel Remy, by Gascoyne's *A Short Survey of Surrealism*, published in 1935 and providing 'a focus for the new energy, putting into plain words, and inscribing within the history of twentieth-century ideas, the various challenges they represented to conservative, smug aesthetic isolationism and, beyond, to the prevailing hidebound conceptions of the life of man' (19). However, Gascoyne also presented arguments that were more in line with the positions held by Jennings and Madge. One of his concluding arguments for the importation of Surrealism is that it would fit in with the native traditions of English literature: 'Shakespeare, Marlowe, Swift, Young, Coleridge, Blake, Beddoes, Lear and Carroll' (Gascoyne 2000: 94). This repeats an argument formulated by Madge as early as 1933, although without Madge's emphasis that the persistence of the English tradition should lead to an English Surrealism in which 'we should rather imitate [the French] example in the general motives and fundamental methods of their working, than in the works themselves' (Madge 1933: 18). In a review, Madge wryly summed up Gascoyne's book: 'Surrealism is now in its academic period' (Madge 1935: 21).

The trouble was that there were two overlapping but competing models of English Surrealism: one aiming to use Surrealist techniques to realise a liberated modern mass Englishness; the other aiming to use the English tradition as a peg on which to hang a Surrealism designed to act as a corrective to England. Eventually, this led Jennings and the communist editor of *Contemporary Poetry and Prose*, Roger Roughton, to demand in April 1937 that the Group disband itself. Gascoyne describes the 'gloriously funny' meeting in his journals, which after much abuse and invocation of Lenin wound up with a compromise solution: 'In the end, R. and J. resigned (much to everyone's relief), and we all went downstairs to drink beer and whiskey' (Gascoyne 1991: 74). It is reasonable to assume that part of the problem had originated with Read's definition of surrealism as 'a reaffirmation of the romantic principle' (28) and his call in its name for 'a revaluation of all aesthetic values' (45) in the introduction to the post Exhibition volume, *Surrealism*, he edited for Faber. His comparison of surrealism to a spontaneous 'international and fraternal *organism*' as opposed to an artificial 'collective *organisation* such as the League of Nations' (20) was insufficient to distance his position from Richard's call for 'a League of Nations for the moral ordering of the impulses' made ten years before. The strength of Jennings's opposition to what he perceived as the movement of English Surrealism away from the metamorphosis of the intellectual stream of images and towards the literary substitution of active poetic images, is reflected in the caustic rhetoric of his review:

> Now a special attachment to certain sides of Surrealism may be defendable, but the elevation of definite 'universal truths of romanticism' (pp. 27–8) in place of the 'universal truths' of classicism is not only a short-sighted horror, but immediately corroborates really grave doubts already existent about the *use* of Surrealism in this country. We all agree with Mr Read that the eternally fabricated 'eternal truths of classicism' constantly appear as the symbols and tools of a classical-military-capitalist-ecclesiastical racket. But then we remember a recent query in a film-paper: 'Is it possible that the business of national education is passing, by default, from the offices of Whitehall to the public relations departments of the great corporations?'
>
> Is it possible that in place of a classical-militarist-capitalist-ecclesiastical racket there has come into being a romantic-cultural-soi-disant co-operative-new uplift racket ready and delighted to use the 'universal truths of romanticism – co-eval with the evolving consciousness of mankind' as symbols and tools for its own ends? Our 'advanced' poster designers, our educational propaganda film-makers, our 'young' professors and 'emancipated' business men – what a gift surrealism is to them when it is preserved in the auras of 'necessity', 'culture' and 'truth' (Jennings 1936: 167–8).

Here, then, is a combined critique of English Surrealism and British Documentary, diagnosing their common symptoms, and grimly prophesying the future utility of their shared techniques to advertising, propaganda and modern-day 'spin'. Against this dystopian vision, Jennings advances a position based on Breton's contribution to *Surrealism*. Breton argues that at a point when Europe was poised to become a furnace, the International Exhibition, by unifying 'in one name the aspirations of the inventive writers and artists of all countries', had demonstrated not just a shared style but 'a new consciousness of life *common to all*' (Breton 1936: 99). Aside from the influences of Marx and Freud, Breton ascribes the formation of this universal Surrealist position to two distinct modes of perception: objective humour and objective chance. The latter resides in being alive to the tendency of certain situations simultaneously belonging 'to the real series and to the ideal series of events' and thus providing a 'vantage-point' for agency (103–4). This is a similar point to the one that Madge made about multiple personality being the necessary condition for simultaneous enchantment and liberation, as can be seen further from Breton's subsequent statement: 'It is only at the approach of the fantastic, at a point where human reason loses its control, that the most profound emotion of the individual has the fullest opportunity to express itself ...' (106). In this context, Breton refers to the Gothic castle in the eighteenth-century novels of Anne Radcliffe, Mathew Lewis and Horace Walpole as being one such vantage-point which has, however, developed into:

> ... a point of fixation so precise that it becomes essential to discover what would be the equivalent for our own period. (Everything leads us to believe that there is no question of it being a factory.) However, Surrealism is still only able to point out the change which has taken place from the period of the 'roman noir' [i.e. Gothic novel] to our time, from the most highly charged emotions of the miraculous *apparition* to the no less disconcerting *coincidence*, and to ask us to allow ourselves to be guided towards the unknown by this newest *promise*, brighter than any other at the present time, isolating it whenever possible from the trivial facts of life (112–13).

It is this passage that Jennings draws on in order to admonish Read for seeking to limit Surrealism to Romanticism: 'It is to cling to the apparition with its special "haunt". It is to look for ghosts only on battlements, and on battlements only for ghosts.' Against this, he suggests not only that coincidences hold the promise of the future but also that they have 'the infinite freedom of appearing anywhere, anytime, to anyone' (Jennings 1936: 168). This potential solution to the problem of helping the masses stranded in the image space find the same vantage-point for agency as the surrealist intelligentsia, ensured the centrality of the concept of the coincidence for the initial phase of M-O.

Empson's Imaginary Solution

A less convincing attempt to reconcile the positions of the intellectuals and the masses features in Madge's review of the keynote speeches from the 1934 Soviet Writers' Congress – published in English as *Problems of Soviet Literature* – in the February 1936 issue of *Left Review*. Ideologically committed to endorsing socialist realism and the specific recommendations of the Congress, Madge nevertheless tries to defend the position of the avant-garde intelligentsia while commenting on the controversial debate about the relevance of Joycean techniques for the proletarian novel: 'is it necessary for the proletarian writer to travel this long route in order to reach the world of socialist realism? No, it is not necessary. For Joyce it was necessary, but not for the proletarian writer'. Likewise, while he notes the demand that the bourgeois class should dissolve itself and merge into the proletariat, his own idea of this 'long-drawn-out historical process' is more dialectical: 'They come to meet their proletarian brothers with the heritage of bourgeois and pre-bourgeois culture which must become incorporated and trans-formed in proletarian culture' (230).

The argument looks tortuous now to a posterity not favourably pre-disposed to socialist realism, but what it indicates on Madge's part is not bad faith so much as a desire to find and maintain a universal position from which to defend a modernity he could articulate quite differently in other contexts. The sense that this quest for the universal vantage-point was a shared pursuit of the time is supported by the review immediately following Madge's: J. Brian Harvey's assessment of Empson's *Some Versions of Pastoral*. From his Marxist point of view, Empson's study of pastoral as a form of 'bourgeois humanism', starting with its introductory essay on proletarian literature, indicates an intellectual 'not yet a Marxist' tearing himself away from his own class and 'the rather starched apron-strings of his tutor and mentor, Dr. I.A. Richards' (Harvey 1936: 231–2). However, Harvey's triumphant conclusion – 'The ideas of pastoral, then, are *not* more permanent, since they would disappear if there were no revolutionary proletariat' – suggests that the nature of his Marxism has been changed by reading Empson. Therefore, it is a more appropriate response than was perhaps intended for a book which was partly concerned with reconfiguring the concepts of proletarian literature and socialist realism as advanced in *Problems of Soviet Literature*.

At the Soviet Congress, Maxim Gorky had demanded a return to the critical realism of the best bourgeois novelists such as Dickens and Balzac as an antidote to the ramblings of modernist social degenerates. Russian novelists should write about the real triumphs of the proletariat in the Soviet Union:

> In our days, vast expanses of land are coloured a single mighty hue. Above the village and the country town looms not the church, but huge buildings of public usage; giant factories glitter with a million panes of glass, while the toy-like heathen churches of ancient times speak to us eloquently of the giftedness of our people as expressed in church architecture. The new landscape that has so sharply changed the aspect of our land has not found a place in literature (Gorky et al 1977: 56).

But what type of literature could reflect such a set of social relations? It is difficult to see in practice how any form of critical realism could fail to interrogate and lay bare the alienated character of such labour held in religious thraldom to the glittering factories. This is precisely what the Soviet leadership did not want. Relocating the factory from the urban world to the rural one is a move that seeks to relate the social relationships of the factory to the natural order in a manner analogous to that by which the agrarian capitalism of eighteenth-century England appropriated the empty allegorical form of court pastoral – or counter-pastoral as Raymond Williams was to define it (see Williams 1985: 13–34) – in order to authenticate itself through theatrical allusion to the primary working activities of country life. Similar considerations prompt Empson to point out in *Some Versions of Pastoral* that 'good proletarian art is usually covert pastoral' and that 'any socialist state with an intelligentsia at the capital that felt itself more cultured than the farmers (which it would do; the arts are produced by overcrowding) could produce it; it is common in present-day Russian films, and a great part of their beauty' (13). However, as this suggests, Empson did not see 'the pastoral process of putting the complex into the simple' purely as a means of hiding real social relationships behind a simple natural order but as a 'trick of thought' central to English literature that can be shown to have operated through a precise historical series (25).

Empson's argument is reminiscent of Benjamin's position in his early essay 'Fate and Character'. Benjamin stated that fate and character are not, as commonly thought, causally linked. Fate is the condition of intelligibility that alienated humanity gained by the adoption of mythic ritual – in contrast to the natural condition of lived experience – and it survives into the modern age in the form of legal statutes, which originally determined humanity's relations to the gods. Therefore, Benjamin argues 'it was not in law but in tragedy that the head of genius lifted itself for the first time from the mist of guilt, for in tragedy demonic fate is breached' (Benjamin 1996: 203). But the reason why tragedy is tragedy, is that the realisation of the 'heroes' that they are as good as god leaves them with no framework with which to build a new order. Against which, the comic hero is never at a loss as to what to do because his comic nature lies in simply letting his character traits unfold. These character traits are the traces of a natural life predating the mythic and show up best in comedy through the fact that we laugh at actions which we would normally condemn according to our reified rational morality.

Almost as though following on from Benjamin, Empson starts by outlining his theory of the double plot, which in its earliest form is the 'comic interlude often in prose between serious verse scenes' (29). Here, 'the clown has the wit of the Unconscious' and complements the conscious doings of the heroes – in effect allowing them to have unconscious motivations without losing their dignity – in a way that affects the response of the audience without necessarily making them consciously aware of the technique. This develops into the double plot proper where two stories are juxtaposed to form a unity. There appears to be a division between highborn, heroic, tragic strands on the one hand and lowborn, common, comic strands on the other. This state is referred to by Empson as the 'device prior to irony' which 'sets your judgement free because you need not identify yourself firmly with any one of the characters; a situation is repeated for quite different characters, and this puts the main interest in the situation not the characters' (50–1).

Thus there is a merging of binaries such that the consciousness of the audience member is not allowed to slip into a simple identification with the characters most obviously representing their social position. Gradually this shifts into a situation where the double plot actually functions for one set of characters in which both sets of possibilities can be seen. This form becomes more overtly ironical as the audience becomes conscious of the double meanings and possibilities. At the same time these versions of pastoral were becoming increasingly oppositional to protestant puritan consciousness and nascent capitalism. This comes into focus in Empson's chapter on John Gay's *The Beggar's Opera* which is subtitled 'Mock-Pastoral as the Cult of Independence'. This functions, as Empson makes explicit, through double irony – a dramatic irony is piled on to the already existing irony present in the double plot structure: 'The trick of style that makes this plausible [and this seems to be the trick of thought that Empson had in mind in the introduction concerning proletarian literature] is Comic Primness, the double irony in the acceptance of a convention.' This is not single critical irony which Empson tells us would be 'I pretend to agree with this only to make you use your judgement and see what is wrong' (170).

Comic Primness can be divided into three categories or levels with respect to the acceptance of social conventions:

(1) Acceptance implies that conventions are right and that it is good to keep them – but unexpected applications serve to show them as superficial in such a manner that the activities of the normal person are unchanged by them. In other words, this is the space in which the everyday traces of natural life can most easily find expression below the ceiling of convention. This is most clearly visible in the world of poverty and tavern life which features in Gay. Empson describes it as the site of '"free" comedy'.

(2) Acceptance implies that the conventions are wrong as in the case of critical irony, but the double irony lies in the further implication that the speaker will nonetheless comply – either through weakness or for personal advancement. Here, conventions, while perceived to be arbitrary, form a comfortable shelter safe from the absolutes of natural life from which to operate. Empson describes this as a site for '"critical" comedy'.

(3) Simultaneous acceptance of and revolt against the convention adopted primly by the speaker. This is the site of '"full" comedy' because the enjoyer gets the joke at both levels. As Empson comments, 'For this pleasure of effective momentary simplification the arguments of the two sides must be pulling their weight on the ironist, and though he might be sincerely indignant if told so it is fair to call him conscious of them. A character who accepts

this way of thinking tends to be forced into isolation by sheer strength of mind, and so into a philosophy of Independence' (171).

This last is the most difficult to understand and Empson provides some examples such as ironical humility. This is where the speaker says something like 'I am not clever, educated, wellborn' and then implies standards so high that those of the other before whom they are humbling themselves must be quite out of sight. This implies valuing the conventions of education and social rank so highly that there is plenty of space left below in which to live freely (first level comic primness) while simultaneously ironically rejecting the same conventions. Without the first part of the manoeuvre this would be merely another example of second level comic primness, but the holding open of the everyday traces of natural life transcends a mere sheltering behind convention and raises the ironist to the third level of comic primness.

This 'philosophy of Independence' achieved by a 'trick of thought' close to simultaneous multiple personality, is very similar to Surrealism's optimum vantage-point for liberation as subsequently outlined by Breton. Not only do Breton and Empson both emphasise concentration on the repetition of situations, but they also both demand what Benjamin described as 'character', Empson as 'full comedy' and Breton as 'objective humour': 'Humour, as a paradoxical triumph of the pleasure principle over real conditions at a moment when they may be considered to be particularly unfavourable, is naturally called upon as a defence during the period heavily loaded with menaces in which we live' (Breton 1936: 103). Breton suggests that the English, with Swift and Lewis Carroll, are better fitted than any to understand this humour, which holds Surrealism in active tension by forming a complementary pole to that of objective chance and the coincidence. Empson notes: 'Alice has, I understand, become a patron saint of the Surrealists, but they do not go in for Comic Primness ...' (221). The advantage of the third level of Comic Primness was that it automatically held simultaneity and humour in tension and therefore eliminated the need for luck which Madge had identified in 1934. It looks very much as though Breton's 1936 essay is moving towards Comic Primness, perhaps in a form mediated by discussions with Jennings during collaboration on the organising committee of the International Exhibition. Therefore, the concluding remarks of Breton's essay, in which he suggests that Surrealism could only gain from contact with the English tradition, might amount to more than polite flattery.

Empson insists on the importance of philosophical independence to the development of a genuine proletarian literature by arguing that Gorky's definition of socialist realism transforming the world, through a process of extracting images from real existence and allowing them to signify in new fields of desire, applies to all good literature. However, as he demonstrates, it is not easy to say what is actually proletarian about proletarian literature:

> One might define proletarian art as the propaganda of a factory-working class which feels its interests opposed to the factory owners'; this narrow sense is perhaps what is usually meant but not very interesting. You couldn't have proletarian literature in this sense in a successful socialist state. The wider sense of the term includes such folk-literature as is by the people, for the people, and about the people. But most fairy stories and ballads, though 'by' and 'for', are not 'about'; whereas pastoral though 'about' is not 'by' or 'for' (13).

It is from this position that he comes to the conclusion that good proletarian art is usually covert pastoral. If the old pastoral showed a beautiful relation between rich and poor, 'the effect was in some degree to combine in the reader or author the merits of the two sorts; he was made to mirror in himself more completely the effective elements of the society he lived in' (17). But this was an unconscious process, or at least had to be treated as unconscious in order to maintain its seriousness and not display the humour of covert-pastoral. Once the process becomes conscious with the advent of mock- or covert-pastoral, the author, and the readers who pick up on this, acquire the necessary philosophical independence: that of being able to think in the two modes simultaneously. Moreover, covert-pastoral enacts a moment of mutual recognition of the Other that does not automatically break down into an Hegelian division between being-for-self and being-for-others. Therefore, Empson criticises the Soviet approach to both proletarian literature and dialectics in general: 'I do not mean to say that the philosophy is wrong; for that matter pastoral is worked from the same philosophical ideas as proletarian literature – the difference is that it brings in the absolute less prematurely' (25).

The problem, however, with this independence, as Empson admits, is that its necessary recognition of difference has historically been most commonly associated with bourgeois individualism. This implication, along with the sexual joke entailed by 'bringing in the absolute less prematurely', is taken advantage of by the dominant model of 'Thirties' reception in its determination to portray Empson as indulging in 'an elaborate critical conceit, an intellectual's joke at the expense of reductive communist critics' (Hynes 1982: 170). By suggesting that 'Empson is only half trying it on' (Cunningham 1988: 141), the joke is made doubly reassuring for not only are the Marxist conceits pricked, but it also triumphantly restates the truths of good bourgeois art:

> It is an ingenious argument, with several important consequences. First, it protects the artist from being 'proletarianised'; he remains the Artist – separate, intellectually superior, a conscious maker. Second, it preserves the idea of Art as artefact, the sum of the artist's tricks. Third, it allows for submerged, 'secret' levels of meaning below the conscious, polemical surface. And fourth, it assumes that there is a relation between excellence and complexity – that is, it makes the standard of judgement aesthetic not political (Hynes 1982: 172).

It may be an ingenious argument but it is not the one that Empson makes in *Some Versions of Pastoral*. He does not elevate the 'Artist' as hero because he recognises that the heroic is merely a special instance of pastoral depending upon who is personified: 'If you choose an important [person] the result is heroic, if you choose an unimportant one it is pastoral' (70). He treats both as equally premature attempts to transform the world and by referring to Christ as 'sacrificial cult-hero' pricks the conceits of Christianity. His main concern with the figure of the 'Worker' in proletarian literature is that it is the same sort of 'mythical cult-figure' as the hero and therefore just as amenable to Tory propaganda as any specifically communist approach (19–20). By designating the 'proletarian artist' Grierson's *Drifters* as pastoral (14–15), he condemns it for the same faults as Renaissance heroic drama. For, as Jennings had pointed out in his letter on 'Triumph', the need was no longer for heroes – whether workers or kings – to save us from nature, but for us to save us from ourselves.

These issues came to a head in the 1930s as a consequence of the divergence of the course followed by the mock-pastoral form since its eighteenth-century heyday. The rise of rational industrial capitalism and its decisive break with rural values banished undisguised mock pastoral to the nursery, either in the form of new works such as *Alice in Wonderland* or through the infantilisation of old ones like *Gulliver's*

Travels. However, concurrently, comic primness continued into modernity as a 'certain tone' which, as Peter Nicholls argues, became the origin of modernism. For example, Baudelaire's 'To a Red-haired Beggar Girl' functions as pastoral: an ostensible celebration of the girl's natural beauty in comparison to finery and artifice. Yet, as Nicholls makes clear, this is highly ambiguous: 'The element of voyeurism here, as the poet clothes and unclothes the girl, is closely bound up with what Baudelaire celebrates elsewhere as the "cold detachment" of the poet-as-dandy' (Nicholls 1995: 2). As Baudelaire is quite prepared to admit that his social position is as wretched as the girl's, the pastoral independence he achieves translates into an aesthetic, rather than a societal, mastery.

This move results in poetic vision no longer being tied to the idea of social transformation as it was in romanticism, because in its pursuit of aesthetic mastery it is rejecting not only the natural but all forms of the social, not least because the prospect of transformation within that sphere appeared increasingly remote in the years after 1848. The consequence of this was a cleavage between this aesthetic modernity and a bourgeois modernity which corresponds to the material belief in progress of the second half of the nineteenth century. Interestingly, these same two categories are called respectively counter-pastoral-modernism and pastoral-modernism by Marshall Berman (Berman 1983: 134). These terms are related to, but not necessarily interchangeable with, the similar terms used by Empson and Williams. More specifically, Berman's pastoral-modernism – a 'faith in the bourgeoisie neglect[ing] all the darker potentialities of its economic and political drives' (135) – corresponds to Williams's counter-pastoral in which capitalism is displayed as the natural order. Other examples include the Soviet attempt to locate the factory in the countryside, Gorky's definition of socialist realism and most advertising copy. These forms all include a transformative social vision, but only at the level of an unconscious understanding: no possibility of independence is generated. Against these forms stands Berman's counter-pastoral-modernism which, following Baudelaire, achieves a certain ironic detachment and aesthetic independence at the cost of abandoning the social sphere: 'the counter-pastoral image of the modern world generates a remarkable pastoral vision of the modern artist who floats, untouched freely above it' (139). Examples of this form would include the work of the high modernists and the Surrealists.

The situation that developed in modernity can therefore be seen as a split between two similar sets of operations that are both versions of pastoral and forms of modernism – although they are more conventionally, and misleadingly, separated into realism and modernism. Looked at in this light, the attempts of someone like Madge to break down the barriers between Surrealism and socialist realism appears perfectly understandable. However, the point for Empson was that both modernism and realism remained premature versions of pastoral. It was only the appearance of proletarian literature following the Russian Revolution that reconnected pastoral with the wider tradition of fairy tales and folk literature and thus reintroduced the natural traits that lay in its origins with the clown, so enabling comic primness in the subversion of social conventions.

The really elaborate critical conceit and intellectual joke being enjoyed by Empson, therefore, was that it is only by subjecting the 'false limitation' (11) of proletarian literature to mockery that a genuinely liberated multiple consciousness can come into being. In effect, Empson offers readers a solution to the 'puzzling form' (13) of pastoral in a manner displaying similarities with the playful 'Note on Notes' in his second volume of poetry, *The Gathering Storm*. He describes these notes as like the answers to a crossword puzzle, arguing that there would be no point in publishing the puzzle if it was so easy that it did not require the answers to be printed, before concluding that 'it is always part of the claim of the puzzle in poetry that this is the best way to say something' (Empson 1940: 55–6). The best way for Empson to show how individuals being forced into a 'philosophy of Independence' fulfils the otherwise unrealisable goal of proletarian literature, is for him to display double irony in the acceptance of literary, social and political conventions. His own third level comic primness serves not to elevate him above the 'simple person' but paradoxically allows him to share the simple virtues – 'luck, freshness, or divine grace' (22–3) – through the complex employment of more than one mode of thought simultaneously. This technique is an extension of that facility for 'popular poetry' which had been identified by Michael Roberts in *New Signatures*, as lying in Empson's complex compression of the analogy between ideas and the analogy between emotional responses. Here, as in Breton's definition of objective chance – the simultaneous participation of certain situations in both the real and the ideal series of events – a precise fruitful ambiguity has the potential to transfigure the world. However, the achievement of *Some Versions of Pastoral* is to shift this potential from being a property of situations to being a property of individual agency. As Terry Eagleton sums up, pastoral provides Empson 'with a kind of imaginary solution to a pressing historical problem: the problem of the intellectual's relation to "common humanity", the relation between a tolerant intellectual scepticism and more taxing convictions, and the social relevance of a professionalized criticism to a crisis-ridden society' (Eagleton 1983: 53).

The question was how could this imaginary solution be put into practice? One possible example which appeared relatively soon after *Some Versions of Pastoral* was Orwell's *The Road to Wigan Pier*. This impudently satirised the conventions of the documentary movement in order to subvert the heroic representation of the 'Worker' with a pastoral vision of the domestic interior:

> I have often been struck by the peculiar easy completeness, the perfect symmetry as it were, of a working class interior at its best. Especially on winter evenings after tea, when the fire glows in the open range and dances mirrored in the steel fender, when Father, in shirt-sleeves, sits in the rocking chair at one side of the fire reading the racing finals, and Mother sits on the other with her sewing, and the children are happy with a pennorth of mint humbugs, and the dog lolls roasting himself on the rag mat – it is a good place to be in, provided that you can be not only in it but sufficiently *of* it to be taken for granted (Orwell 1998: 107–8).

This passage, which was widely influential on postwar British represen- tations of the working class, presciently evokes Trebitsch's definition of Henri Lefebvre's conception of everyday life: 'both a parody of lost plenitude and the last remaining vestige of that plenitude' (Trebitsch 1991: xxiv). Orwell's half-parody is written in the knowledge that the culture it describes, and the trace of everyday plenitude at its heart, are under threat from the 'Utopian future' of 'rubber, glass and steel', where 'everyone is educated', dogs and horses 'have been suppressed' and 'there won't be so many children, either, if the birth-controllers have their way' (Orwell 1998: 108–9).

We can see the description of Orwell's working-class interior as com- plying with the first level of comic primness: a celebration of a life freely lived underneath conventions. Opposed to this is the insincerity of middle-class socialists – characteristic of the second level of comic primness – which Orwell attacks in the form of 'the outer-suburban creeping Jesus ... who goes about saying "Why must we level *down*? Why not level *up*?" and proposes to level the working class "up" (up to his own standard) by means of hygiene, fruit-juice, birth-control, poetry etc' (150). Orwell tries to show this grouping that they, themselves, are part of the wider process in which the whole middle class is becoming part of a mass mechanised society: 'It is only romantic fools who flatter themselves that they have escaped, like the literary gent in his Tudor cottage with bathroom h and c, and the he-man who goes off to live a primitive life in the jungle with a Mannlicher rifle and four wagon-loads of tinned food' (203). To this end he offers his own public persona as a representative model for an ironic accommodation with modern life in which '...we have nothing to lose but our aitches' (215). However, it is precisely this final qualification which raises Orwell's comic primness from the second to the third level as he simultane- ously accepts and revolts against middle-class conventions, while holding open the everyday plenitude of the working-class interior – the promise of 'a good place to be in' – against the coming 'bleak world'.

At the time of writing *Wigan Pier*, Orwell was linked with the group of writers around John Middleton Murry's *Adelphi* and close to Jack Common in particular. Common displayed a working-class comic primness by writing his own articles under the by-line of 'The Sweeper Up'. Both accepting that he was part of the mass and simultaneously revolting against the idea of the mass, Common won Orwell's admira- tion by directing his efforts at trying to change the nature of the mass rather than trying to escape from it through his highbrow literary con- nections. His perspective dovetailed with that of Orwell, coming together at the same problem from a different angle, and his influence can be seen in the argument of *Wigan Pier* that working-class ex- perience has to be a key element of the cultural consciousness of the new mass society advocated by Orwell.

Common's significance to the development of mass society in the long twentieth century is as one of the first working-class intellectuals to argue that traditional working-class culture had no future. He under- stood that the skilled working class, which had roots stretching back into artisanal traditions, was being bypassed by the changes in the state and society following the First World War, so that unskilled but collectively organised workers such as the railwaymen – popularly referred to as the 'pease-pudding men' because 'things were dolloped out to them soft as pease-pudding on a paper' – were becoming what might be termed the new labour aristocracy. These were the workers who could afford mortgages or rents on the new housing estates and consumer goods such as radiograms, who developed educated cultural interests and encouraged their children to get on at schools. As such they were potentially part of the 'new lower middle class' constellation identified by Jeffery as forming the core of the M-O volunteers and, in fact, he particularly singles out two railwaymen in his account (Jeffery 1978: 30). However, Common, writing in 1935 from his unique vantage-point of proletarian at the *Adelphi*, opens up a new dimension for examining the social condition of this class:

> All this is just enough to give bourgeois apologists some reason for saying that the working-classes of this country have become bourgeois .... The pease-pudding man is now safely ensconced in his five-roomed semi-detached, his garden is about him, the world's classics of music and literature are on his shelves, his job is a permanent one reasonably likely to be well-paid. He is, you might say, in the position of the moderately comfortable petty-bourgeois; he might set out as his predecessors in the class above him did, to impose his views and habits on the world about him. He doesn't, though. There is the catch. His present position depends on his maintaining his proletarian loyalties and yet delimiting them. Let him play the individualist game in authentic bourgeois fashion, and he's down. It is power of striking and standing by his mates, not his enterprise, which keeps him afloat. On the other hand, if he permits class loyalty to run away with him, he must unite with the unemployed and the unions of the starved crafts: it means sharing their poverty sooner or later. To do that successfully means finding a communal formula which will be a true crystallisation of the proletarian ethic. That's what is needed, a second crystallisation, trade unionism being the first (Common 1980: 39).

The sense that Common is describing his own dilemma as much as that of the pease-pudding men gives the passage a strength which emphasises the urgency of the need for 'a cultural consciousness which squares with [working-class] communal experience' (Common 1980: 41). As such it reinforces the similar arguments of middle-class intellectuals such as Kracauer, Benjamin and Orwell. All were agreed that a celebration of the plenitude of proletarian existence as an end in itself was a form of retreat from the real contradictions of everyday life. The problem was to find a comparable point of agency in the space created by the structural transformation and separation of bourgeois and proletarian consciousness. The surrealists advocated various techniques generating a vantage-point through the creation of simultaneous multiple personalities but these remained heavily dependent upon chance. Empson's discussion of the possibility of 'proletarian literature' in the form of covert pastoral offers, as Eagleton suggests, an 'imaginary solution' to this problem in the form of comic primness which guarantees a 'philosophy of Independence'. Moreover, the adoption of similar tactics by writers such as Orwell and Common indicates that this is more than an imaginary solution at the individual level. Common, in particular, offers a model for a new self-consciousness rooted simultaneously in the senses of being an individual and a part of a larger collective whole still in touch with the animal functions of a natural life: he was both intellectual and member of the mass. The next chapter examines how M-O set about attempting to replicate such an Empsonian solution at the mass level.

4
Early Mass-Observation

The Formation of Mass-Observation

In 1976, Charles Madge described the immediate context of 'The Birth of Mass-Observation' in the *Times Literary Supplement*:

> As a reporter on the *Daily Mirror*, I was in 1936 one of many journalists helping to 'cover' the events leading up to the abdication of Edward VIII. Deployed now here, now there by my news editor, I stood little chance of an overall view of what was going on, but at least what I did know was at first hand, and of potentially more historical interest than the largely fabricated and contradictory accounts that appeared in the newspapers, including my own (1395).

The press had completely covered up the developing constitutional crisis that had begun with the realisation that Mrs Simpson was seeking a divorce in October 1936 and then when the story did finally break at the beginning of December, it misrepresented developments so much that the actual Abdication came as a distressing shock to many of the public. The press and the public were both split over the issue and the fact that Winston Churchill – always a viable alternative Prime Minister as history was to show – publically supported the King against his own party's Government led to a level of tension in which rumours of an impending Spanish-style civil war began to circulate. Madge later recalled the zenith of his reporting career: 'On December 10, the day of the abdication, I was sent to intercept Winston Churchill on his return from seeing the King, carrying a blank cheque which I was to offer him in return for his exclusive story of this meeting' (Madge 1987: 64). Churchill declined.

Both the myths surrounding the event and subsequent historical accounts serve to obscure the fundamental division in British society that was laid bare by the crisis. The National Government, led by Stanley Baldwin, resented the modern king's popularity and tendency to publicly comment on slums and unemployment. Therefore, they were quick to take advantage of the situation created by Edward's desire to marry Mrs Simpson and forced the constitutional crisis by refusing to countenance a marriage which would allow her to be his wife but not queen. The king could potentially have sat out the crisis waiting for Mrs Simpson's decree *nisi* to be made absolute, but a statement was lodged by a law clerk in a firm sometimes employed by Baldwin that if proved would have derailed the process. As an eminent historian has observed: 'Remorselessly, the king was driven towards abdication' (Taylor 1970: 493). It was this conflict between modernity and tradition which, rather than the supposed gaps between the press and the people and the leaders and the led that became a frequent lament, was to become the faultline running through M-O. In *The Long Weekend*, a social history of Britain in the interwar period first published in 1940, the poet and novelist Robert Graves and the former Mass-Observer Alan Hodge describe the support of the ordinary people for the King: 'In London crowds packed Downing Street, chanting 'We want our King', and at Woolworth's the Edward VIII Coronation mugs were rapidly sold out' (364). In the twenty-first century, it is difficult to appreciate the social significance accorded to the Woolworth's-going public in the 1930s. However, to give one telling example, in September 1934 fifty delegates at a conference called by the ailing British publishing industry unanimously concluded that a 'New Reading Public' existed but that only Woolworth's and the tuppenny libraries were catering for it (L. Jones 1985: 13). It was this state of affairs that led Allen Lane to target Penguin books at Woolworth's and the new reading public when he launched them the following year. Trial sales of the initial ten titles published by Penguin proved so successful that Woolworth's placed a consignment order for 63,500 more copies and Penguin, which had been seen as a wild gamble, broke even and changed the face of British publishing. Penguin's most successful venture was to be the thirty six political Penguin Specials published before the war, averaging sales of 100,000 copies per title, and radicalising a generation (see Hubble 2003: 107–10). Among these Penguin Specials was M-O's *Britain by Mass-Observation*, appearing in January 1939. M-O, Penguin and the new reading public who patronised Woolworth's can all be seen as elements of the classless society that Orwell was to identify in *The Lion and the*

Unicorn. Therefore, it is not overly stretching the point to suggest that the Abdication crisis marked the opening salvoes of what was to become the 'People's War': the wartime struggle between the National Government and an extraparliamentary coalition of progressive forces such as the Commonwealth Party, J.B. Priestley's 1941 Committee, M-O and individuals like Orwell.

Madge shared his concern over the failure of the press to report what people actually thought about the abdication with Jennings and the other friends who had already begun to meet at Blackheath. This led to their thinking about 'the possibility of enlisting volunteers for the observation both of social happenings like the Abdication and also of "everyday life," as lived by themselves and those around them' (Madge 1976: 1395). In December 1936, the group – which included Madge, Raine, Jennings, Cicely Jennings, Legg, Gascoyne, Hunter and Ruthven Todd (see N. Stanley 1981: 6) – prepared a questionnaire, which gives us some idea of the scope of these initial discussions:

Name
Address
 1 Age
 2 Married or unmarried
 3 What are your superstitions, in order of importance?
 4 Do you pay any attention to coincidences?
 5 What is your class?
 6 What is your Father's profession and your own?
 7 Do you or did you hate your Father, and if so, why?
 8 Do you or did you hate your Mother, and if so, why?
 9 Do you or did you want to get away from home, and if so, why?
10 Do you want to have a son, or a daughter, or both?
11 Do you hate your boss: do you hate your job?
12 What is your greatest ambition?
13 Did you want the King to marry Mrs Simpson, and if so, why?
14 Were you glad or sorry when the Crystal Palace was burnt down, and if so, why?
15 Do you approve of the institution of marriage as it exists in this country at present? If not, how would you wish it changed?
16 Are you in favour of the disestablishment of the Church of England?
17 Are you religious? If so, in what form?
18 Do you welcome or shrink from the contact by touch or smell of your fellow men?

19 Can you believe you are going to die?
20 How do you want to die?
21 What are you most frightened of?
22 What do you mean by freedom? (MOA: FR A4).

Among the accompanying notes are several instructions which make it clear that the questionnaire was supposed to be administered to third parties: 'Answers should be obtained from the person questioned at a speed which will prevent him from taking refuge in a merely conventional and socially correct response.' The fact that the questioner is referred to as the 'observer' and the third party as the 'subject' suggests an initial position of observation of the masses rather than by the masses. However, most significantly, a comment to the effect that new queries should be substituted for questions 13 and 14 when these become no longer topical, casts light on the exact interest in the role of the King: 'If possible, they should refer to symbolic events, such as the burning of the Crystal Palace, rather than to the real crises of which these events are symbols.' This clearly reflects the influence of *The Golden Bough* and implies similar thinking to Jennings's letter to Empson on 'Renaissance Triumph'.

In the same month, a letter to the *New Statesman* from Geoffrey Pyke raised the issues of whether the press was responding to, or moulding, public opinion about what he described as the 'sexual situation' (i.e. the prospect of Mrs Simpson becoming Queen). Pyke pointed out that, as the response of primitive tribes to sexual situations was one of the main interests of anthropologists, the contemporary situation provided 'some of the material for that anthropological study of our own civilisation of which we stand in such desperate need' (Pyke 1936). This served as the stimulus for Madge to reply, in a letter published by the *New Statesman* on 2 January 1937, that such a study was already underway:

English anthropology, however, hitherto identified with 'folk-lore,' has to deal with elements so repressed that only what is admitted to be a first-class upheaval brings them to the surface. Such was the threatened marriage to the new 'Father-of-the-people' to Mrs. Ernest Simpson. Fieldwork, i.e., the collection of evidence of mass wish-situations, has otherwise to proceed in a far more roundabout way than the anthropologist has been accustomed to in Africa or Australia. Clues to this situation may turn up in the popular phenomenon of the 'coincidence' (Madge 1937a).

This repeated mention of 'coincidences', contemporary with Jennings's review of *Surrealism*, highlights the importance of the concept in this initial phase. Madge mentions this in his autobiography: 'Neither Humphrey nor I were inclined toward Jungian ideas of a collective unconscious, but we had read Freud's essay on the coincidence, which had led to an interest among certain French surrealists, especially André Breton, in coincidental happenings of various kinds' (Madge 1987: 64). Aside from his essay in the *Surrealism* book, Madge and Jennings would also have been aware of Breton's exposition of the 'Political Position of Surrealism' – in response to the Soviet Writers' Congress – as being in line with Marx's demand to create 'more aware-ness' (Breton 1972: 229). Breton argues that in a situation where society and the Left in particular have run out of energy tantalisingly close to the threshold of a new society, it is necessary to create a 'col-lective myth' (210, 231–2). Among the requirements for which are that 'poetry must be created by everyone' (262) and 'the organisation of perceptions with an objective tendency around subjective elements' (278). It is this model that we see in the idea of 'Popular Poetry' and in Jennings and Madge's early M-O article for *New Verse*, 'Poetic Des-cription and Mass-Observation', where they explain that through M-O what is subjective becomes 'objective because the subjectivity of the observer is one of the facts under observation'. Therefore, 'what has become unnoticed through familiarity is raised into consciousness again' (3).

It appears that the Freud essay they were interested in is chapter twelve of *Psychopathology of Everyday Life*, 'Determinism, Chance, and Superstitious Beliefs'. Here Freud discusses the unconscious recog-nition of unconscious processes as manifested in the phenomena of paranoia and superstition. Superstition is a specialised – frequently col-lective – form of paranoia in which unconscious fears are projected on to chance external events. Freud sees the opportunity arising from the sudden experience of a superstitious moment in oneself, or a coinci-dence or a feeling of *déjà vu* or some such sensation, as being that unconscious activity is brought to surface consciousness and can be analysed. It was exactly these unconsious fears and desires that M-O hoped to recover at a mass rather than individual level in the public response to the chance event of the burning-down of the Crystal Palace, which was symbolic of the abdication crisis.

Madge's letter in the *New Statesman* attracted the attention of Harrisson, reading in Bolton public library, owing to the coincidence that it was immediately adjacent to his poem 'Coconut Moon'. He wrote to Madge and met him, Jennings and the others on an occasion celebrated in an oft-quoted passage from Gascoyne's journals: 'what I chiefly remember of the evening is the picture of Humphrey, with his elbow on one end of the mantelpiece, and Harrisson, with *his* elbow on the other end of the mantelpiece, both talking loudly and simultaneously to those present in general, without either of them paying the slightest attention to what the other was saying' (Gascoyne 1980: 10). *Savage Civilisation*, finished the previous summer, had been published by Gollancz on 11 January 1937 to an instant success that took it through several printings before the end of the year, including a Left Book Club edition. However, despite a number of other prestigious activities since his return from Malekula, including giving a well-received lecture to the Royal Geographical Society, writing a series of articles for the *News Chronicle* and appearing on radio and television through the services of Mary Adams, Harrisson had been living in Bolton since the autumn of 1936. Ostensibly there to investigate the link to Malekula via the Unilever combine, Harrisson was also taking refuge from the wreckage of his personal life: the relationship with his father was in the final stages of deterioration and Zita was divorcing Baker to marry Crossman (see Heimann 1998: 113–18, 123–7). Once again, he was only able to resolve the tension between his intellectual and everyday needs through the therapeutic activity of participant observation:

> At various stages I got jobs as lorry driver, shop assistant, labourer, cotton operative, ice-cream man and reporter in firms in or to do with Unilever. The evenings, necessarily sprinkled with eau de cologne, I sat at the fireside of the prosperous Lever relatives, feeling slightly guilty but softly elated. For the first discovery that I made (for myself) in Bolton, was that it came just as easy to penetrate other kinds of western society, as societies in which you are from the start in 'stranger situation' (Harrisson 1959: 159).

The result of the January meeting with Harrisson was another letter to the *New Statesman*, appearing on 30 January 1937, in the name of 'Mass Observation' (as yet unhyphenated) and signed by Harrisson, Jennings and Madge. Now, given that Madge and Jennings had already been working on ideas revolving around Surrealism and PP for some months at least and that there was to be a sharp division in 1937 between their work on day-diaries and Harrisson's organisation of the Worktown project – both of which operations were already in motion to some extent before the meeting – it is tempting to assume that M-O was fatally fissured from the start, as seems to be born out by the subsequent acrimonious dispute between Madge and Harrisson in 1940. However, as we have seen, the three co-founders had a number of things in common such as social and educational backgrounds and a shared anti-colonialism. More importantly, they all strongly held a similar outlook concerning the need to recognise the social and psychological changes occurring in the twentieth century and the consequent importance of finding a way out of the everyday struggles of individual predicament by promoting a new mass social consciousness. Unfortunately, more critical energy has been expended on highlighting the superficial differences than on the important work of investigating why this particular time proved such a fruitful historical conjuncture for a groundbreaking interdisciplinary social project. For example, the personal statements of Madge and Harrisson in the booklet *Mass-Observation* illustrate the problems of trying to construe their relationship as a simple opposition between interpretation and positivism respectively:

> Tom Harrisson believes that Mass-Observation by laying open to doubt all existing philosophies of life as possibly incomplete, yet by refusing to neglect the significance of any of them, may make a new synthesis ... The whole study should cause us to reassess our inflated opinions of our progress and culture, altering our judgements on others accordingly ....
>
> In the other author's opinion, Mass-Observation is an instrument for collecting facts, not a means for producing a synthetic philosophy, a super-science or super-politics ... It is each man's job to find his own salvation as best he can. Mass-Observation merely proposes to acquaint him with relevant scientific facts (Madge and Harrisson 1937: 47–8).

Yet there has been a tendency for critics simply to reverse these statements surreptitiously in order to suit their purposes: 'One strange thing about these two statements is that one would have expected the latter statement to have been written by Tom Harrisson, even if Charles Madge could never have written the first. Harrisson was always the scientist, he was determined to collect facts ...' (Jeffery 1978: 23).

In fact, the joining of former expedition leader Harrisson with former political organiser Madge meant that there were now two people with experience in the planning and running of group activities, even before the film and theatre experience of Jennings was taken into consideration. At a practical level, the relative independence of Madge and Harrisson with respect to each other was to make the whole operation possible. Madge ran the national projects from London while Harrisson ran the Worktown project in Bolton. Each knew what he wanted to do and both had the connections to produce published results. Through Eliot, Madge was able to get Faber and Faber to commission *May the Twelfth*; while from Gollancz, Harrisson received advances for four books – of which only one was ever to see the light of day and that not until 1943 – with which he was able to employ full-time observers. Thus, in the beginning there was no need to argue over resources and no real opportunity for friction to occur in the day-to-day running of the organisation.

In the immediate enthusiasm following the founding of M-O, Harrisson and Madge sat down together to write what was to become the introductory booklet *Mass-Observation* and completed the first two sections before the end of January (9–28). The theme throughout these pages is a detailed engagement with superstition, which they argue has its roots in 'the earliest days of prehistoric man' and is 'infinitely adaptable' – something which had become ever more apparent with the onset of modernity:

> The more recent acquisitions – electricity, the aeroplane, the radio – are so new that the process of adaption to them is still going on. It is within the scope of the science of Mass-Observation to watch the process taking place – perhaps to play some part in determining the adaption of old superstitions to new conditions. These forces are so new and so terrific that they are commonly thought of as kinds of magic power that can only be wielded by a few men, the technicians. Hence there is a widespread fatalism among the mass about present and future effects of science, and a tendency to leave them alone as beyond the scope of the intervention of the common man. The technician on the other hand, is not concerned with the implications of his activity or its effect on the masses (16).

The situation is portrayed as a Wellsian choice, in the age of gas and the bomber, between scientific and anti-scientific attitudes, which Madge and Harrison reduce to a simple question: 'which gives us most hope of surviving?' Taking the answer for granted, they discuss how the mass can be educated scientifically. Historically the problem had been that while the industrial revolution created the conditions enabling mass education and literacy, it 'had a disruptive effect on the morals and beliefs of the working class' throwing up two strands – effectively, respectable and tavern – equally resistant to science: 'It is only gradually that the prolongation of industrialism has made more numerous a third type who has looked for a solution in terms of science rather than of religion' (18). This suggests Common and Orwell's arguments about the ongoing developments by which part of the working class were becoming part of the new middle class. In particular, the emphasis on the figure of the technician anticipates Orwell's argument in *The Lion and the Unicorn* that it is technicians, mechanics, radio experts, industrial chemists, film producers, popular journalists and higher-paid skilled workers 'who are most at home in and most definitely *of* the modern world' (Orwell 2000b: 408). However, while Harrisson and Madge place more emphasis on a rational attitude to gender equality, it is clear that the 'second crystallisation' they were seeking was more overtly scientific than the cultural consciousness sought by Common and Orwell.

This scientific emphasis is explicitly emphasised in a political context as a necessary response to one of the major consequences of the structural transformation of society: 'the availability of the entire mass, through its literacy, to suggestion for commercial, political or other reasons' (18–19). The situation in 1937 with advertisers and newspapers employing 'the best empirical anthropologists and psychologists of the country' to 'aim their suggestions at the part of the human mind in which the superstitious elements predominate' (20) called out for mass science. Already, we can see how the relationship to superstition is changing from the initial surrealist-inspired approach. It is no longer seen as an opportunity to bring the unconscious to the surface for analysis, but as something to be identified and combated in the name of science. However, what is actually meant by science remains ambiguous. On the one hand, we are told that this science is subsuming art, which flourished previously only because it could attempt to answer the questions that science was not yet ready for (25–6). Yet, at the same time, we are informed: 'In certain branches of science and of art, the individual scientist or artist becomes absorbed in a collective activity which is purely human in type, and which excludes neither of the two categories (27).

It is not clear if the science proposed is to be an all-encompassing method that must be adhered to, or something to be built up from the close observation of everyday life. Nor can this dichotomy be assigned to the divide between Madge and Harrisson, as Madge's article 'Magic and Materialism' published in *Left Review* that February makes clear. Here, he repeats the same two positions albeit with different nuances. He acknowledges that science will supersede poetry but in a way that emphasises poetry's freedom from superstition: 'Poetry deals, not with the inexplicable, but with what has not yet been explained. It lights up by fitful flashes, a scene on which the full day of science will presently dawn' (32). At the same time, he tries to reconcile socialist science with socialist realism:

> The wishes and needs of mankind are rendered accessible, on a class-basis, to the artist-scientist, but the nature of his field of inquiry, as scientist, or his subject-matter, as artist, is found to extend beyond himself as observer. His observations must be mass-observations, his data mass-data. His works of art must satisfy not his own isolated fantasy, but the needs and wishes of the masses; his scientific generalisations must apply not only to himself but to every member of his society. His problem is not to raise to the level of his own consciousness aspects of humanity hitherto concealed or only guessed at, but he has to raise the level of consciousness collectively of the whole mass, he has to induce self-realisation on a mass scale (33).

These positions suggest an equal overall ambiguity to that expressed in *Mass-Observation*.

However, 'Magic and Materialism' also represents an explicit attempt by Madge to work through his own poetic and scientific positions with reference to Darwin, Frazer, Freud and, particularly, Marx. Hence the article starts with that passage from the *Communist Manifesto* which has become the most succinct definition of the experience of modernity: 'All that is solid melts into air, all that is holy is profaned, and man is at last compelled to face with sober senses his real conditions of life and his relations with his kind.' Madge argues that humanity can only face its real conditions 'when the proletariat has asserted its materialism over the human sphere' (32). On the one hand, he establishes a series of analogies between poetry, magic and the pleasure principle as all being immature stages on the way to an objective and scientific adaption to reality. While, on the other hand, he notes that modern-day science extends only over the realms of conscious knowledge leaving anything beyond that unexplained and, therefore, liable to a reactionary superstition which seriously inhibits the establishment of proletarian materialism. Moreover, the very social changes characteristic of modernity fostered the rise of fresh superstitions as newspapers and advertising, new forms relying on the old magical power of words over things, pandered to the pleasure principle. In a footnote, Madge suggests that these apparently opposed trends are not as irreconcilable as they appear: 'Perhaps man will return to his instinctive materialism, which has now become transformed into the conscious materialism of science, after an intervening period of magical thought, the period of the *unconscious*' (32n). Taken with what we know of Madge's intellectual background, this implies that a combination of poetic imagery and surrealistic techniques will make the unconscious conscious and free humanity from the symbolic order. However, he was writing with his *Left Review* hand and so, in a similar manner to his review of *Problems of Soviet Literature*, he apparently subordinates modernism to socialist realism in the main body of his text even while the techniques of the former are clearly necessary to the logic of his argument that art and science will become united when both the artist and the scientist – this latter in the figure of the anthropologist – become concerned with both the outer world and the inner self.

Paul C. Ray, drawing on Raine's *Defending Ancient Springs*, describes M-O as 'surrealism in reverse' because instead of projecting the imagination on to the world in order to change it, it tried 'to recover the imagination that produced the vulgar objects and images of the everyday world' (Ray 1971: 177–8). Yet Madge's writing shows that M-O attempted to carry out both operations simultaneously and it is Raine's account in *The Land Unknown* which best captures the calm intensity of the time:

On the wall of the Jennings' room in Blackheath ... hung a painting by Magritte [*On the Threshold of Liberty*, which Jennings owned at the time]. In the foreground a cannon, emblem of coming war or revolution, was pointed towards a wall or flimsy screen, partitioned into sections. I ought to remember them all, for Heaven knows I gazed at that picture often enough. These were the fragments of a world, not, like Eliot's, 'shored against our ruins', but to be demolished when the cannon fired. There was a section of the fac[,]ade of a house; a woman's torso; the trunks of some trees; a patch of blue sky with white clouds ... For all our intoxicating sense of undisclosed marvels under the thin surface of consciousness, we yet saw in that gun pointed at the flimsy fabric of a painted scene the true emblem of the future of our world. It did not dismay us; that is how the spirit of Revolution wanted it to be; the cannon, now about to fire, was our will' (84–5).

However, the career trajectory of Madge from 1935 to 1945, which can be broadly summarised as passing from an earlier poetic position through the mass science position and on into the heart of planning the post-war state, shows that neither this state of imaginative exaltation nor the dialectical synthesis of 'Magic and Materialism' lasted indefinitely. Nonetheless, Madge's writings from the first two or three years of M-O demonstrate the persistence of this synthesis before its eventual collapse. Rather than viewed only as a quirk of individual psychology, this can be seen as registering perhaps the most significant feature of M-O: that its methodology and organisational structure combined to hold open a unique dialectical perspective over a relatively prolonged period of time.

The reasons for this start to become clear from the third section of *Mass-Observation*, written by the end of April 1937, concerning what the organisation is actually doing and intending to do. Here, they discuss the fundamental importance of presenting results so that both academicism and 'the facile temptations of popular exposition' are avoided and science reaches the mass in the form of 'the completely objective fact' (39–40). The actual examples of presentation methods given suggest that this concept of the 'fact' is not as straightforward as it appears. Firstly, there remains a distinct aesthetic component that promises to bring a new perception to 'even the drab and sordid features of industrial life' indicating an approach more akin to Orwell's newly published *Wigan Pier* than the mainstream documentary tradition: 'His squalid boarding-house will become for the observer what the entrails of the dog-fish are to the zoologist – the material of science and the source of its *divina voluptas*' (29–30). Secondly, the metaphor of the detective, suggestive of the discovery of clues and traces, comes with a surrealist twist: 'In the detection which we intend to practise, there is no criminal and all human beings are of equal interest' (30). Finally, it emerges that the object of detection will be the image, 'something between an idea and a sensation', for which purpose the observers will be trained: 'We intend to issue series of images, like packs of playing cards, and to suggest various exercises which can be played with them' (37–8). The idea was to spread the expertise of painters and poets in the use of imagery to the mass-observers, thus liberating their perceptions from externally imposed sense associations and creating a sense of the possibility of change.

The importance of this third section of *Mass-Observation* is brought out by the preface to *May the Twelfth*, dated August 1937: 'The main development of Mass-Observation has been two-fold, firstly the net-work of Observers all over the country; secondly an intensive survey of a single town. Charles Madge runs the former, Tom Harrisson the latter. Humphrey Jennings is responsible for the business of presenting results. These three activities are closely linked' (Jennings and Madge 1937b: iv). This tripartite operational division is common to both accounts and it is how M-O functioned from January to September of 1937. Thus, it can be seen that there is a potential division in the text of *Mass-Observation* between the ambiguously poised mass science position of Madge and Harrisson, and the operational structure which also included Jennings. However, in practice the tripartite structure held the mass science position at its moment of equilibrium, because with Jennings presenting results there was no question of the poetic, image-based technique collapsing into an all-encompassing scientific narra-tive. When Jennings left M-O, this tripartite structure simply ceased to exist and the equally originating mass science position of Madge and Harrisson was released from the tensions that had held it constantly at a point of fruitful equilibrium and consigned to a more unstable hit and miss existence, which still nevertheless maintained many of the original qualities.

However, the initial nine months or so were heady days indeed. Harrisson's friend from Malekula, the Australian Jock Marshall, arrived in England on 14 February 1937 and was shortly confiding to his diary: 'Most people are fairly normal here, but the crowd of young intellectuals which Tom seems to lead have a curious sort of generally understandable at the moment exaggerated kind of conversation of inconsequentialities – none of which have any real logic – oh hateful word – or method ... Catchword at the moment, "esoteric synthesis"' (cited in Heimann 1998: 134). It has been easy to portray M-O as the work of dilettantes of the patrician class, but they could lay a perfectly reasonable claim to be attempting to initiate the mass into the secrets of their proposed synthe-sis. Furthermore, the inner circle were by no means all upper-middle class. Raine's autobiography bears testament to her early fears that on leaving Cambridge she might have to return to her parents at Ilford and become a schoolteacher or a 'city clerk's wife' (Raine 1975: 57). Admittedly, she was later riven by guilt because of her 'commoness', both for attracting Madge with it and for inflicting it on his mother (78–80), and it was 'flight' from the mass (81) which would eventually take her away from both Madge and M-O. This attitude caused her to write of another member of the Blackheath group at the core of M-O: 'There is nothing, in David Gascoyne's kind and quality of imagination, which is typical of, expressive of, suburban values or modes of thought; he is no more of the world into which he was born than the angel Tolstoy's cobbler found naked in the snow behind a church, and brought into his house to learn shoe-making' (Raine 1967: 35). Yet the poetry and example of both Gascoyne, the son of a bank clerk, and Raine herself helped lead the way to precisely that condition of postwar Britain which she bemoaned: 'Many, even most writers of the present time (and among them poets) have come from lower middle-class suburbs; talent is at home anywhere, and numbers of writers have described suburban life in terms of suburban or working-class values; have made those values articulate, comprehensible, acceptable' (Raine 1967: 35).

Another early participant was Dennis Chapman, a graduate from the LSE, who was subsequently to participate in the Worktown economics project, join the WSS and become a member of the University of Liverpool's Sociology department after the war. He was present at a group meeting with Madge following 'the first return of "observers"' contributions after the *New Statesman's* invitation', but it is not clear if he had previously attended. (Chapman 2002). Chapman not only came from a working class and trade union background (Chapman 1979b: 9) but also had social research experience on a project investigating the effects of long-term and juvenile unemployment among Dundee jute workers, which had begun in October 1935. This was led by Oscar Oeser of the Psychology Department at the University of St Andrews and is listed – alongside the Insitute for Social Research (the Frankfurt School) amongst others – under the short section on 'Organisations with Similar Aims', compiled by Hunter, which concludes *Mass-Observation*. Oeser's project was innovative in its interdisciplinary combination of the methods of cultural anthropology, sociology and economics in the name of a scientific social psychology (Oeser 1937: 344). In a self-declared aim to gain deeper psychological insight than social surveys such as *Middletown*, Oeser emphasised the importance of 'functional penetration' – a concept partly taken from the famous Austrian study of the unemployed at Marienthal – as a fieldwork technique: 'observers approach the community to be studied not as reporters with notebook and camera, but as far as possible as accepted members of that community, having several definite and easily intelligible functions within it' (352). As Liz Stanley points out, Oeser's description of his own work as 'anthropology at home' suggests 'an idea that had found its moment rather than being the particular invention of M-O, Malinowski or both' (L. Stanley 1990: 20). Nick Stanley suggests that Oeser's April 1937 article, 'Methods and Assumptions of Field Work in Social Psychology', outlines a template that was adopted by M-O and that Madge and Harrison's failure to acknowledge Oeser's help with the M-O Blackpool survey of July 1937 indicates a deliberate downplaying of his influence – a suggestion which Madge denied in an interview with Stanley (N. Stanley 1981: 49–50). The fact is that M-O had planned most of their programme before Oeser's article was published and though there are similarities in approach, it is doubtful if M-O's practice in Worktown could be described as functional penetration – in any case the examples Oeser gives are hardly inspiring: 'making collections of stamps and clothes, rent collecting for social settlements, giving lectures in unemployed clubs, and so on' (354). Moreover, Oeser's study remained firmly in the British tradition of investigating what it took for a 'social problem' (362), utilising intelligence testing (354) and identifying a disorientated but resistant 'subculture' (358) – none of which became standard M-O approaches.

There was one aspect of M-O's agenda of which we can be certain that it was totally original and this was launched in the joint letter of 30 January 1937 with the promise that 'observers will ... provide the points from which can be plotted weather-maps of public feeling in a crisis' (Harrison, Jennings and Madge 1937: 155). Weather maps are a means of plotting non-linear sequences – chaos theory originated from research into the weather – thus they provide a means of countering the linearity of schedule and calendar time while remaining relatively simple to understand and to act on. M-O wanted to plot popular feeling on a day to day basis:

> Every day the social consciousness is modified by the news reported in the newspapers and on the wireless. The more exciting the news, the more unified does the social consciousness become in its absorption with a single theme. The abdication of King Edward VIII was a focusing point of that kind. The coronation of King George VI is providing another (Madge and Harrisson 1937: 30).

So, while Harrisson organised the intensive study of Bolton, Jennings and Madge instigated day-surveys to take place on the twelfth of each month, in which the observers – recruited from the *New Statesman* letters and sympathetic articles in the popular press – recorded all that they did and saw on that day. These were an end in themselves and ran throughout 1937, but the first three in February, March and April were also trial runs for that to take place on 12 May 1937, the date set for the coronation of George VI.

There is no question that the day-survey idea was innovative and exciting and lengthy extracts were published in the printed monthly bulletins that M-O would start to circulate to observers from August, making good on its promise to publish results. The cover of *Mass-Observation*, designed by Jennings, consisting of extracts detailing the weather first thing in the morning, literally typifies the refreshing everyday perspective promoted by the surveys. However, the 'barometer-readings' from which the metaphorical weather-map would be constructed were to be taken from each observer's dominant image of the day, recorded as per the following instructions:

> The observer is to ask himself at the end of each day what image has been dominant in it. This image should, if possible, be one which has forced itself on him and which has confirmed its importance by recurrence of some kind. The image may occur in a series of varying forms or may take the form of a coincidence. For example, the same name or object may forcibly strike the observer's notice from within or without, several times on the same day (Madge 1937b: 34).

Not only would M-O be able to tell if these images were typical of the day, but also if they were typical of a certain area or class. These images had more than one role. In a mass science sense they were 'completely objective facts', in that they would comprise neither academic nor journalistic report, nor function as the points of reference in a controlling discourse. Moreover, in the modernist sense, images would present both their own meaning and the potentiality of its transformation into other meanings. Furthermore, according to Jennings, they were held in a framework of ritual relationships with an ancient cosmological symbolic order that had shaped beliefs for centuries. The mixture of a normal day-survey and an abnormal coronation day-survey provided the perfect match of experiment and control experiment to fully investigate the power of the image. Therefore, it is not surprising that Jennings and Madge put so much effort into studying and presenting the observations they collected from March 12 and May 12 and had ready for publication by September as *May the Twelfth: Mass-Observation Day-Surveys 1937*.

Coronation Pastoral

Mass-Observation began with a newspaper headline: 'THE KING WANTS TO MARRY MRS. SIMPSON: CABINET ADVISES "NO"'. And followed it with Frazer's definition of the primitive king: 'He lives hedged in by ceremonious etiquette, a network of prohibitions and observances, of which the intention is not to contribute to his dignity, much less to his comfort, but to restrain him from conduct which, by disturbing the harmony of nature, might involve himself, his people and the universe in one common catastrophe' (9). Coronations were a special instance of that ceremonious etiquette, a form of ritual that equated the replacement of the old king by a new king with a symbolic combat of nature such as that between night and day. By being performed smoothly, the ritual could ensure the smooth continuation of the natural order. Yet M-O had come into being because the abdication crisis had unexpectedly caused a breakdown in ritual and therefore in the symbolic order itself: 'Millions of people who passed their lives as the obedient automata of a system now had to make a personal choice, almost for the first time since birth' (9). There had been a period when no official ritual existed to direct the collective image space, at which juncture it was possible that an alternative configuration might arise rendering restoration of the previous symbolic order impossible. At the very least, though, it might be expected that the temporary freedom from order would allow new images to appear which, in turn, might cause the symbolic order to be altered in various ways. Therefore, M-O's unwritten questions with respect to the coronation were: in what manner will the symbolic order be restored? And how can any emergent elements be prevented from disappearing in this process?

The equally unwritten answers to these questions are to be found in the 400 pages of *May the Twelfth*. The book is in five parts; the first four of which are textual montages. The first – press clippings from the build-up to the coronation – and the second section – the events surrounding the procession in London – were edited by Jennings. The third and fourth sections concerned events happening on that day around the country and individual reactions respectively. A fifth section consisting of complete and abridged day-surveys from March 12 in succession was edited by Madge. Other helpers with the editorial process, as listed on the title page, were T.O. Beachcroft, head of the Unilever advertising agency (Madge 1978b: 5) and a short story writer who had contributed to *New Country*, Julian Blackburn, a social psychology lecturer at the LSE, Empson, Legg and Raine. The poet Ruthven Todd compiled the index.

It is important to remember that Jennings and Madge were not just avant-garde intellectuals but also, at the time of the founding of M-O, workers with day jobs in the media as documentary film maker and newspaper reporter respectively. As such, they belonged to that list of technically-minded workers that Orwell was to describe as being most at home in the modern world. Therefore, alongside its scientific and aesthetic contexts, the book is also presented as though to a modern technically-aware audience:

> In addition to the ordinary difficulties of scientific presentation, the editors of this volume have had to consider difficulties of other kinds – difficulties comparable to those which the film industry and the newspaper industry have been built up to overcome. These are the difficulties of appealing to a public of unknown size and of every kind, and the elementary financial problems arising out of this. But again, MASS-OBSERVATION is more than journalism or film documentary, because it has the aim of classifying and analysing, the immediate human world (Jennings and Madge 1937b: 414).

The introduction to the key London section explains how the three types of observer report – the forty three day-surveys, CO.1–CO.43; various responses to a questionnaire, CL.1–CL.109; and the 'Mobile Squad' in touch with M-O headquarters by telephone, CM.1–CM.12 – were arranged to ensure that 'close-up and long-shot, detail and ensemble, were all provided' (90). However, close-ups are a double-edged technique as Benjamin describes in his essay 'The Work of Art in the Age of Mechanical Reproduction'. The technical reproducibility of art and nature through photography and film has the effect of robbing the original object of its authenticity by coverting it into the form of the endlessly exchangeable image. In this process, the role of the close-up is crucial in satisfying a desire to bring the object into the reach of humanity – in the form of magazine pictures, for instance – and so negate it. Everyday perception, itself, becomes so attuned to these processes that it automatically extracts even unique objects from the sphere of 'tradition'. Yet, Benjamin – contrary to some received opinion – did not mourn the loss of this tradition but regarded its shattering as simultaneously marking the reverse of 'the present crisis and [the] renewal of humanity' (215). For the close-up also explodes the world of offices and factories and makes possible their reconfiguration: 'With the close-up, space expands ... the enlargement of a snapshot does not simply render more precise what in any case was visible, though unclear: it reveals entirely new structural formations of the object' (229–30).

His thinking in these areas makes it clear that Benjamin would have understood and supported the ideas behind M-O: 'The adjustment of reality to the masses and of the masses to reality is a process of unlimited scope, as much for thinking as for perception' (217). Indeed, Benjamin goes further than the founders themselves in expressing the logic of M-O, by arguing that mass audiences are replacing the function of the critic. The critic occupies a position of intense contemplation before the work of art and so remains within the sphere of tradition and, therefore, cannot be free of the ritualistic element which ties it to the symbolic order. However, the new everyday perception – or, rather, apperception – of the mass audience is a dislocating awareness of the new spatial and temporal possibilities, opened up by technical processes and the general structural transformation of society, that ruptures the possibility of the kind of contemplation displayed by the critic and opens up 'an immense and unexpected field of action' (229). Exposure to dislocation, in the traffic filled streets of the modern city as well as in film, leads to a heightened presence of mind more attuned to becoming conscious of its apperceptive experience. This could be developed by a committed – committed firstly in the technical sense – art using close-up and montage to encourage the imagination by the mutation of space into social forms capable of embodying a new lived experience. To paraphrase Benjamin's statement in the essay on Eduard Fuchs, technology has given the masses the strength to shake the burden of accumulated treasures off its back, so as to get its hands on them (see Benjamin 2002: 268).

However, the danger was that the same forces of dislocation could be used to strengthen tradition by intensifying imagery linked to the symbolic order in the manner of fascist rallies and radio and film propaganda: 'The violation of the masses, whom Fascism, with its *Führer* cult, forces to their knees, has its counterpart in the violation of an apparatus which is pressed into the production of ritual values' (Benjamin 1992: 234). Film, especially newsreels, enabled anyone and everyone to have a role in the ceremonial rituals which had once been the exclusive right of kings and priests. The masses were brought face to face with themselves in 'big parades and monster rallies, in sports events, and in war' (234n). War, in particular, offered the ultimate walk-on part; the only means by which the masses could express themselves and all the technical production processes of society could be maintained in full, without altering property, and hence class, relations. This is what Benjamin understood as the aestheticisation of politics characteristic of Fascism. The only response was to politicise art.

This danger that the masses be submerged within the total spectacle is recognised by *May the Twelfth* in both the 'revival of a wartime atmosphere' – obviously added to by the presence of large numbers of troops mobilised to line the procession – and the 'curious record-hunting ascetic feeling' revealed in the desire of people to queue through the night in order to be part of the proceedings (91–2). The text refuses complicity in these processes by means of its technical composition. This is accomplished by intercutting a series of examples of the crowd reducing the threat of the military by their comments and behaviour – 'There goes the Salvation Army'; 'Where' ave they 'oused all these soldiers?' (footnote in text: 'In spoken English you would more often talk of "housing" an emu or an elephant; the idea that soldiers are pet animals seems to crop up here'); 'feeding the troops' by throwing sweets at them (139, 141, 147) – with a series of analogies that deflate the idea of taking part in a momentous occasion: 'Some of these people look as though they were going to a funeral not a Coronation'; 'I heard one of my neighbours remark that she thought it was like dog-racing – something to see for a short time and then nothing for considerably longer'; '...the route is lined with soldiers who are standing in front of a mass of people who look like refugees from Guernica'; 'One damned thing after another' (117, 121, 122, 140).

But more subversively, the text switches the focus entirely onto the crowd and, therefore, generates a stance of mass comic primness towards the monarchy – supposedly at the centre of attention – represented only in the dislocated form of glimpsed coaches. In keeping with this Empsonian influence, the account of the actual day in London is prefaced with the entire soliloquy from *Henry V*, IV, i (87–8). Empson discusses the character of Henry in *Some Versions of Pastoral*, arguing that 'there is something fishy about him'. The point is that 'the Henries are usurpers; however great the virtues of Henry V may be, however rightly the nation may glory in his deeds', Shakespeare has a 'double attitude' to him (87). The soliloquy demonstrates Henry's ironic acceptance of the conventions of kingship – thus placing him at Empson's second level of comic primness:

O Ceremony, show me but thy worth.
What! is thy soul of adoration?
Art thou aught else but place, degree and form,

Creating awe and fear in other men?
Wherin thou art less happy, being fear'd,
Than they in fearing.

But conversely he is well aware of the free pleasures of living life beneath the ceiling of conventions and thus he really displays the third level of comic primness:

And but for ceremony, such a wretch,
Winding up days with toil, and nights with sleep,
Had the fore-hand and vantage of a King.
The slave, a member of the country's peace,
Enjoys it; but in gross brain little wots;
What watch the King keeps, to maintain the peace;
Whose hours the peasant best advantages.

Henry's consciousness of being forced into philosophical independence, and his consequent mastery, derive from his knowledge of himself as playing a part. Shakespeare's own double attitude allows him to pose as critic or admirer depending on the audience's viewpoint, but always safe from the charge of sedition, while at the same time proclaiming his own superior independence in the egotistical manner of renaissance humanism.

Following this, the description of the scenes in London on Coronation day proceed like the renaissance double plot of heroic and pastoral – only with the heroic half missing. We see nothing of the king: the nearest the heroic gets to our attention is as a series of confused identities reflected in snatches of vigorous argument from the street lining crowd:

Woman: 'No, not yet. I think the Duchess first.'
Another: 'That's the Queen.' Then, disappointed: 'No.'
'That's Queen Mary.'
'That's Princess Marina.'
'Princess Royal, that is.'
'The Queen of Norway.'
'I saw Marina.'
'I'm sure Queen Mary's next.'
Man shouts: 'Hullo, George, boy. Well, Marina.'
Woman: 'This is Queen Mary's coach next.'
(It is evident that no one in the crowd actually knows who is who)
(134).

Hundreds of years of existence as the half-erased traces of a history written by the victors is overturned by a 'camera' shift. This is illustrated by Jennings's own photographs taken on coronation day, which show jumbles of heads, legs and feet; against backgrounds of litter, leaves, London Underground arches, steps, scaffolding of stands, trees against the skyline, statues and lampposts (not included in *May the Twelfth*; see instead M.L. Jennings 1982 or Jackson 2004). Together, these form an exploration of the possible metamorphoses of public space which pays no attention to the king and stands as a direct challenge to a more conventional 'news reel' style depiction of a large flag-waving crowd perfectly foregrounding the important personages at the front – such as the photograph unfortunately chosen for the cover of the fiftieth year anniversary reissue of the book.

This ambiguity – the possibility of either type of representation – is encapsulated in M-O's declaration that what is being presented is a 'panorama of London, and especially of the route of the Coronation procession' (91). The panorama was originally one of the means by which landscape was brought closer to humanity by technical reproduction. The landscape is ripped out of its function as a material reference point and can either be liberated as public space or mythicised as a set of interchangeable units within a symbolic totality. Within the modern techniques of conventional mass media representation, the coronation procession, itself the remnant of a mythic hierarchy now secularised to situate humanity in a totality embodied in the symbolic public figure of the king, had the capability of inducing a previously unachievable total participation in the event. One way this was done in practice, despite the actual physical progress of the procession following a scheduled route, was through a system of loud-speakers relaying the cheering of the crowd from the point actually being passed by the procession around the rest of the route and so creating a dislocating shock triggering involuntary participation (138n). This is felt by the observer who reports 'the most stirring incident was the unreasonably (so it seemed) fervent cheering I felt compelled to give with others to the King and Queen on their return' (128). It is more strikingly shown in the account of another observer who says of her young daughter: 'When the show was over I found Lydia [age 5] still croaking dreary and monotonous cheers until I stopped her' (131).

May the Twelfth fractures this temporally-induced dislocated participation *temporally* by appealing beyond dislocation to the heightened apperceptive awareness of mass modernity. We do not follow the procession, we follow a chain of observers around the route reporting on overlapping time spans. We do not perceive the disconcerting relayed roar of the crowd as a continuous burst of manifested total being; but as a series of relayed roars, a series of relayed injunctions not to throw streamers out of the windows, a series of repetitions from some other distant plane of existence that seem no more than a faintly ridiculous officiousness when compared with the immediate experience of beer drinking, singing, tree climbing and courting (137–9). The technical effects of a surrealist-inspired montage have been used to create a narrative irony overlaying the irony of the Coronation double plot.

However, neither M-O or *May the Twelfth* had managed to 'have the effect of lessening the magical power of the symbol of the monarchy' (Madge 1937b: 33). Furthermore, there are fundamental questions as to what M-O's carnivalesque demonstration of the momentary independence of the masses achieved. It could be argued that this type of independence only ever exists through the social relationship with the monarch: that the King's pure consciousness of being-for-self, and hence his authority to rule, is guaranteed precisely by the independence of the masses's being-for-others, and *vice versa*. Therefore, *May the Twelfth* holds the independence of the coronation crowds open at exactly the moment when its public manifestation was required in order to legitimate the new King; and in reality that moment was just converted back into information in the form of radio and television broadcasts – to an estimated 60,000 viewers as far away as Brighton, Ipswich and Cambridge (284n) – and film reels flown round the country to be shown that very evening (286–7). Everyday life resumed and the public crisis over the abdication came to an end. The resistant plenitude identified was not ultimately a threat to the social order: it was the underpinning of that order.

Such a reading, however, would be blind to the strategies adopted by Jennings, who crucially was both in the text as one of the observers (CM1) and operating at a level removed as the editor. The independence of the masses depicted in the sequence is not a consequence of their social relationship with the King, but a consequence of their textual relationship with Jennings. The technique is analogous to that of *Wigan Pier*, with Jennings using the interplay between his textual and editorial personas, to transfer his artistic independence into a collective independence – thus fulfilling the need, as outlined by Madge in 'Magic and Materialism', for the artist to move beyond his own desires in order to satisfy the wishes of the masses. To these ends Jennings employs a similar aesthetic form of half-parody to Orwell's: 'At Hyde Park Corner Rovers are hurriedly putting up a metal barrier in the centre of the street where a lot of cardboard boxes (left by periscope and chocolate sellers) are lying on the ground in the rain. They are now as slippery as banana peels. A girl is lying on the ground in the arms of a policeman' (144). It is the same mixture of a parody of lost plenitude and the last remaining vestiges of that plenitude, compiled in the same knowledge that the moment described is not going to last: 'The open stands are empty. The statue of Byron shines in the rain. The police are reforming their units' (145). These images – and Jennings's photography – are like crime scenes posing a challenge to the viewer (see Benjamin 1992: 219–20). Consequently, rather than the effect which has been ascribed to the book of 'putting the reader there as though she or he were watching from the top of a slow-moving bus' (Pocock 1987: 419) – which conveniently ignores the fact that there was a bus strike on that day – we are invited to become detectives in a case that can never be closed: an unrelenting investigation of everyday life 'at once a rejection of the inauthentic and the alienated, and an unearthing of the human which still lies buried therein' (Trebitsch 1991: xxiv). Shortly after completing this work, Jennings left M-O to work as a free lance film director (Jackson 2004: 206).

By mapping the moments of crisis and resolution, M-O had fulfilled at least part of their original intentions by bringing some of the unconscious mechanisms of everyday life to light. Madge used these to formulate a tentative theory of society as initially outlined in the leaflet, 'A Thousand Mass-Observers':

The main study of Mass-Observation at present is in fact *the impact of society on the individual*. 'Society' is an abstract word but it represents concretely to every single person a whole number of other people who affect his life. These people are of three types – they fall into three areas. Nearest home, in what we term Area One, are his (or her) family, the people he sees every day in his home or lodging and at his place of work. Then comes Area 2: meetings with strangers. Outermost, but of peculiar power in influencing his life, is Area 3, filled by all those names of celebrities and public figures, film-stars, footballers, kings, mythical heroes, characters in news, history and fiction; people like Gracie Fields and Earl Baldwin, whom he may only see or hear at secondhand, on the wireless, on the screen, in the newspaper, in books; people who govern him, affect his actions and the way he parts his hair ... M-O is studying the shifting relations between the individual and these three areas, thus seeking to give a more concrete meaning to the abstract word, 'society' (M-O 1937a: 6–7).

As this leaflet also includes an extract from an observer's report on 12 July, it seems reasonable to assume that it was published around the end of that month. The M-O bulletin for August – the first of a series of printed monthly bulletins which ran regularly until the following January and more sporadically thereafter – refers to the 'Aug. 12' day-survey, reminding observers that the 'guiding principle is that of the three social areas', and outlines how the theory will be used to investigate social habits such as smoking (M-O A: FR A4). The results of this investigation were subsequently reported in *First Year's Work* (10–12). However, the most extensive published analysis employing the theory is the fifth section of *May the Twelfth* (347–414). Here, it can be deduced that the concept of the three social areas – illustrated by a diagram of three concentric circles (348) – is, as Empson would say, worked from the same philosophical ideas as comic primness. Area 1 is the everyday part of life not governed by social conventions. Area 2 is the public sphere where conventions do hold sway. Area 3 is the media realm of leaders and celebrities, which appears to have been constructed, as a footnote in *May the Twelfth* suggests, through an exploration of the analogy between the relationship between the king and his subjects and that of the modern relationship between the media and the masses:

> The King is the archetype of all the personages in area 3. On the great public occasion of his Coronation he exhibits himself in the flesh to his subjects. This is obviously of the greatest importance as a means of establishing his position at the centre of the entire social system. Hence it is that broadcasting plays so vital a role, in enabling the contact between areas 1 and 3 to be effected on a far wider scale than has ever been possible hitherto (14n).

For M-O, the independence of the King, like the independence of Shakespeare's Henry V, is the independence of Empson's third level of comic primness: the ability to accept and revolt against convention at the same time. Monarchy is only one possible expression of this independence and no longer the model for all other relationships in society. However, the very fact of its anachronistic survival allows the media's role in determining societal relationships to be made visible. The dilution of the dialectical nature of the relationship between the king and the people to just one moment among a chain of differences demonstrates the ability of the media to bridge the gap between areas 1 and 3, thereby eroding the role of area 2. By analysing the contradictions so momentarily visible, M-O were unearthing changes akin to what Habermas later called the structural transformation of the public sphere.

Unfortunately, this achievement is much easier to discern today than it was for contemporary critics. By and large, reviews of *May the Twelfth* focused on the coronation scenes and ignored the final section. However, two years later, the anthropologist Raymond Firth published an extremely critical review article concerning the history of M-O to that date, which included a particular withering account of the three social areas:

> Sixty pages of material from the 'normal-day survey' are reproduced with notes, putting items into areas (i), (ii), (iii), with certain subdivisions. The aim appears to be to obtain a classification of all the everyday activities of a given observer and so to define the varying range of his social horizon. After this classification the authors state, 'in concluding this section, it is fitting to point out that any explanations or hypotheses put forward in the course of analysis are the first tentative approach to a new set of scientific problems' [413]. It is difficult to discuss the value of this approach since the generalisations based upon it so far are somewhat sparse. When, for instance, a young man attends a dance with a young woman to whom he is attracted and feels uneasy when she is appropriated by other men to whom she seems to be well known, the penetrating analysis offered is 'she seemed to be in area i for other men, and he wanted her to be in his area i. Again, when a man saw another in a car whom he thought he knew, and waved to him but afterwards realised he didn't know him although the other waved back, he was amused. This is classified by Mass-Observation as 'confusion of areas i and ii.' I leave psychologists and sociologists to contemplate the vistas of new scientific problems opened up by this. To an anthropologist accustomed to the analysis of behaviour in terms of relationships on the basis of kinship, economics, residence, political organisation, and other institutional alignments, this method of concentric circle classification is too subtle (Firth 1939: 190).

Firstly, Firth was missing the point. Secondly, M-O had moved on from this stage of analysis by 1939 as he well knew. Thirdly, the great irony of the situation is that Madge had taken some of his ideas for analysis, as well as his defence that it was all a kind of work in progress, from an anthropological work: Bateson's *Naven*.

One imaginative way of reading the last section of *May the Twelfth* is to interpret the concept of the 'social incident' (371) as simultaneously linking to the emphasis on situation in the ideas of both Empson *and* Bateson. On the one hand, as we have seen, the reporting of repeated incidents throughout *May the Twelfth* enabled the focus to be placed on situation, which not only allowed those conscious of contexts and conventions to express their independence, but also expressed the independence of the masses. On the other hand, M-O's attempt to classify incidents can be seen as a preliminary attempt to map out the social ethos and eidos of the modern mass society in which they lived. By trying to combine both aims, Madge was attempting to fulfil the role he had already described: the artist-scientist, or more specifically the artist-anthropologist. The problem is that it is difficult to gauge his success in this attempt precisely because of the lack of published, and indeed, unpublished sources. Madge came to view this line of work as a failure and the section of *First Year's Work* in which it was written up was extensively redrafted prior to publication. However, as also noted in *First Year's Work*, the monthly day-surveys from 12 February 1937 to 12 January 1938 contained 'approximately 2,300,000 words' and analysis required 'still at least another year's work to be done' (47). This work was never done and so the approach can never be fully assessed.

In any case, in following up his interest in anthropology, Madge began to attend Malinowski's seminar at the LSE during the autumn of 1937 and consequently his position began to change. The influence of Bateson can be seen to be diminishing as he is first described as 'a young and somewhat speculative anthropologist' in the draft version of *First Year's Work* (M-O 1937b: 4) and then omitted from the published version altogether. At the same time, Madge was devising a qualified conception of the artist-anthropologist in which a public stance of scientifically respectable anthropological orthodoxy would be complemented by a private individualist poetic independence. A developed version of this position can be seen in an undated – probably late 1938 from external evidence – draft essay, 'Poetry, Time and Place'. The essay revolves around two lines from Marvell's 'The Garden':

Annihilating all that's made
To a green thought in a green shade.

Therefore, it can be read as a comment on Empson's discussion of the same lines in *Some Versions of Pastoral*. Empson argues that the poem contrasts and reconciles 'intuitive and intellectual modes of apprehension' – in other words, its existence subverts Richards's arguments about active and intellectual streams of poetic images. The lines can be read alternatively as suggesting the reduction of the material world to a thought or considering the material world to be of no value compared with the thought: 'This combines the idea of the conscious mind, including everything because understanding it, and that of the unconscious animal nature, including everything because in harmony with it' (99). As Empson suggests, here magic and materialism are combined in a state of Buddhist-like enlightenment neither conscious nor unconscious – an individualist version of the synthesis Madge was describing in 'Magic and Materialism'.

Madge begins 'Poetry, Time and Place' by referring to Freud's argument in 'Beyond the Pleasure Principle' that time and space do not hold in the unconscious. Madge suggests that the implication of this is that the unconscious mind must be purely qualitative and that, therefore, although one can be conscious of qualities, one cannot be conscious of what qualities are. So while the mind can register simple qualities, like 'green' or 'tree', and derives pleasure from such recognition, special skills have to be acquired in order to recognise qualities too complex for simple words or straightforward formulae to act as their referents. Poets, he argues, are one group of people who have these skills in a developed form. From this perspective, he argues that Marvell's lines show a reconciliation between pleasure and reality principle. That thought can become green suggests the mind can make or unmake the environment, thus reflecting the requirement of the unobstructed pleasure principle that thought should be omnipotent. However, this cannot be a complete surrender to the pleasure principle because if so the reality principle would ensure that 'the conscious mind would refuse to entertain the picture so presented, and the pleasurable effect of the lines would be lost'. The character in the poem has not reverted to the pleasure principle: 'He is repeating that situation on a conscious level' (11). In other words, Madge is referring to an adult extension of that infantile process, which Freud describes in 'Beyond the Pleasure Principle', of establishing mastery through repetition:

This kind of philosophic thinking is a satisfaction of the primitive wish for potent thought which the reality principle does not veto, because experience has shown that such thought is powerful and can lead to alterations in reality. Experience also shows, however, that philosophers may delude themselves, may imagine that they are thinking at the level of the reality principle, and actually be reverting to the primitive level of the pleasure principle (11–12).

There is no suggestion that Madge is discussing the issue as more than a general critical issue and yet, rather like Empson's general critical discussions, it would be unwise to rule out there being conscious implications for the contemporary situation. While this statement is by no means a rejection of the possibility of independent philosophical thinking advanced in various forms by Empson, the surrealists and poetry in general, it does shift the focus from the productiveness to the ambiguity of such positions and mark a qualification of this aspect of the founding position of M-O. More specifically, Madge implies that this poetic philosophical independence is more usefully employed in a private sense, by discussing the example of Marvell's situation as representative of a new kind – originating in the seventeenth century – of individualism: 'the right of the individual to develop his personal sensibility and to speculate in an essentially anarchist way.' Marvell's virtue was that, rather than use this position to exclusively claim the privilege of 'super-sage, with special intuitions about reality', he 'treats "being a poet" as a role which he can adopt at will when off duty in the public sense' from his politically active public role – at one time as a member of parliament (21).

Madge concludes that 'The Garden' works, and has applicability to his contemporary period, because it registers leisure as no longer an exclusive feature of the aristocracy. Therefore, it can be seen to mark the historical origins of the bourgeois public sphere, when private independence began to allow critical and rational intervention in public and political matters. Typically, this private independence is seen as having resided in property rights, but here the possibility is raised of basing public rights on the independence of thought of the type that M-O in both its mass science and popular poetry guises was trying to promote. Madge's move from poet-scientist to private poet and public scientist was, therefore, not so much a retreat from an epiphanic millennial vision as a considered decision to bring in the absolute less prematurely. As such, he anticipated not only the structural transformation of the public sphere, but also Habermas's project of reinstating a critically-debating public sphere in which private expertise would underwrite public intervention. However, in Madge's case, the vision was to take the form of a mass sphere legitimated by individual poetic and philosophical independence.

5
Britain Begins at Home

First Year's Work

Among the 'Worktown Papers' at the M-O A, there is a draft account of 'Bolton Through the Ages' which ends with the following note and quotation:

> Disraeli spent some time observing round Worktown. In *Sybil* he wrote '"Two nations; between whom there is no intercourse and no sympathy; who are ignorant of each others' habits, thoughts and feelings, as if they were dwellers in different zones, or inhabitants of different planets; who are formed by a different breeding, are fed by a different food, are ordered by different manners and are not governed by the same laws."
>
> "You speak of ..." said Egremont, hesitatingly.
>
> "THE RICH AND THE POOR."' (M-O 1938a)

M-O's Worktown project has never quite broken free of nineteenth-century associations such as philanthropy, charity and social exploration, with their suggestions of upper- and middle-class paternalism and voyeurism. This sense has been reinforced by accounts of Harrisson's tendency to invite friends and fellow intellectuals to visit, as Julian Trevelyan recounts:

> Tom asked me up to Bolton first of all to paint it, and I set out with him from London one night at ten o'clock to drive there, with Bill Empson in the back. Along the great trunk road we drove, past the streams of lorries, their lights flashing and dipping in the secret language known only to their drivers. At last, about five in the

morning, Bill became restless, and we had to stop in a café, so that he should not see the dawn that upset him strangely (Trevelyan 1957: 83).

However, the Conradian feel of this epic journey was only half the story of what rapidly turned on arrival at the Worktown headquarters at Davenport Street into a fairy tale: 'A house like any other in Bolton, it contained a few beds and office desks, and an old crone who cooked us bacon and eggs and tea on a smoky grate.' This was a topsy-turvy world in which all the pre-existing hierarchies were reversed. Trevelyan found the novelist John Sommerfield writing about pubs, Empson being dispatched to study the contents of sweet-shop windows and himself making collages of the Bolton streetscape from newspaper reports of the coronation (83–4). One inhabitant of Bolton commented on the collage: 'This gives me the worker versus royalty feeling' (88). It also highlighted the dialectical relationship revealed by *May the Twelfth*, but instead of focusing on the independence of the king and of the mobile squad of observers – notably Jennings himself – to reveal the independence of the masses, it focused on the materiality of industrial Bolton and its resident workforce and, thus, revealed the materiality of the monarchy and the M-O intellectuals. This was to be the literally revolutionary achievement of the Worktown project.

In other respects, though, it remained an unfulfilled project. The announcement of the forthcoming publication of four volumes in *First Year's Work* serves to demonstrate both the scope and the shortfall of the Worktown study:

The first results of the 'Northtown' [as Bolton is referred to throughout the book] survey will be published (Gollancz) in the autumn, thus:

'The Pub and the People,' by J. Sommerfield and Bruce Watkin.

'How Religion Works,' by J.W. Wilcock, Reynold Bray and Derek Kahn.

'Politics and the Non-Voter,' by Walter Hood and Frank Cawson, with Eric Bennet, S. Cramp and B. Barefoot.

These volumes will deal with every phase in three major institutions of a great industrial town (24).

The Northtown survey has been correlated with a parallel study of Blackpool. The results will make a full volume, by H. Howarth and R. Glew (45).

None of these were to appear in the autumn of 1938. Sometime in 1939, the above-mentioned Brian Barefoot wrote an extensive memoir of his work at Bolton – the fullest account of the first phase of the Worktown Survey between 1937 and 1938 – including the information that the publication dates for the books had been postponed to the spring and then to the summer of 1939 (201). He brought his account up to date with a handwritten note to the effect that he had spent ten days in July 1939 finishing chapters of the politics book and that it and the others – with the exception of the religion volume – were ready for the printer at the end of August 'but the war broke out and laid low all Mass-Observation's plans'. The list of M-O publications at the end of *Britain Revisited* includes *Politics and the Non Voter* under 1940, but specifies 'unpublished: in proof' (Harrisson 1961: 269). However, there is no longer any trace of this proof copy. Although *The Pub and the People* – which along with its principal writer, John Sommerfield, will be discussed in chapter seven – subsequently appeared in 1943, none of the other books ever saw the light of day. Nor, incidentally, did any of the 'new cycle of detailed factual studies' – comprised of volumes on 'Social Factors in Economics', 'Men, Women and Sex' and 'The English Day' [the day-survey material] – announced in the M-O Bulletin for December 1938. Therefore, it can be argued that if it had not been for the outbreak of war, M-O would have published substantial amounts of material, thus fulfilling their initial promise and rendering their subsequent erasure from the history of British sociology impossible.

The research for these books was conducted by intensive participant observation on a daily basis over long hours, as described by Barefoot:

> It was full-time work; starting immediately after breakfast, about ten, and working till one, dinner time; then we carried on till tea at five, then after tea till ten, eleven or twelve (some of then worked till 2 a.m., and so did I on occasions), broken only by a supper of fish and chips from the local fish shop down the street (Barefoot 1939: 3).

Barefoot, himself, first worked in Bolton during the summer of 1937 before going to Edinburgh to study medicine. However, he returned at regular intervals – a weekend in November and then longer spells in January, July and August 1938 – to continue working on the project. As a university student, he fits the unfair stereotype that 'a steady if transient supply of unemployed, largely Southern, and university-educated youth willingly took jobs or lived on subsistence pay as "full-time observers" in Bolton' (Cross 1990: 2). While this could be said of some of his fellow workers, this was never the rule as can be seen from an analysis of those listed above as responsible for the proposed publications. Reynold Bray was Harrisson's friend from Oxford in the early 1930s. Derek Kahn is mentioned in *Letter to Oxford* as an Oxford undergraduate (87) and had gone on to become a contributor to *Left Review*, for which he intelligently reviewed Orwell's *Wigan Pier* (Kahn 1937: 186–7). Herbert Howarth and Bruce Watkin were Oxford undergraduates, originally recruited through a visit to the Oxford English Club by Madge in early 1937. Together with Alan Hodge, they had been members of the 'Mobile Squad' at the coronation. They were probably contributors to the 'Oxford Collective Poem' (see Cunningham 1988: 339) and demonstrated the influence of Madge and Jennings by adopting a similar prose montage style as demonstrated in a wonderful joint description of Blackpool Tower, part of a larger piece co-written with Hector Thompson:

> It is a natural erection, which dominates its surroundings for twenty miles in space and three generations in time. The sky is brighter in Blackpool and heaven is nearer. Glass-horse trams reflect the local beauties in their glass-house panels. Sodomy supplies Blackpool with radiation both internal and external, and day by day it is love that keeps the Tower standing (Howarth, Thompson and Watkin 1937).

In a letter to Madge, accompanying this piece, the Lancastrian Howarth explains the methodology employed to compose the piece and notes that his relations to Blackpool are Area 1. Hence, Howarth's leading role in directing field work in Blackpool and editing the material for publication was based on his local knowledge rather than his educational background. His co-editor, Richard Glew, was also a local contact (see Cross 1990: 6).

Like Howarth, Frank Cawson was also Oxford-educated but a native Lancastrian. He first worked at Bolton in the summer of 1937 after finishing teacher training in Liverpool and then, after a term at Middlesborough Grammar School, he returned for the first four months of 1938. Much of Cawson's work was taken up with the West Houghton byelection, which M-O covered in depth by getting Cawson to work for the Conservatives while Walter Hood worked for Labour. The immediate benefit of this operation was that M-O had access to all the canvassing returns from both sides. However, Cawson learnt several weeks afterwards that Harrisson had been selling the Conservative information to the Labour Party and consequently resigned from M-O (Cawson 1980: 4). Cawson later admitted that he did have fears at the time that he was 'slumming it' and suspected that this was certainly true of some of the others who visited Bolton but not of the principals like Harrisson, Madge and Sommerfield, nor of his fellow 'foot soldiers' like Hood. Indeed, like Barefoot, Cawson's attitude was that he had been one of the real pioneers of M-O as against the later recruits who were mainly 'clerk-types' and 'yes-men' (8).

Walter Hood and Joe Willcock, both working-class 'foot soldiers', were in Bolton from the original leasing of the Davenport Street house in March 1937. Willcock had been a tramp preacher and the warden of the St Christopher Hostel in the East End of London, where Harrisson had spent several of his school holidays helping out (see Barefoot 1939: 70; Green 1970: 105). His account of the problems entailed in 'signing on' at the Labour Exchange is one of the highlights of *Speak for Yourself* (Calder and Sheridan 1984: 23–8). Hood, who was to become the Commonwealth officer of the TUC after the war, came from a Durham mining community and worked as a miner for three years. Politically-active and an accomplished orator, he toured the country speaking for the Labour movement in a van funded by Stafford Cripps, having rejected the personal overtures of Harry Pollitt to join the CP. Eventually he followed the working-class intellectual route of going on a scholarship to Ruskin College, where he met Trevelyan and Coldstream through his artwork. Leaving Ruskin in 1936 at the age of thirty, he went to live with Trevelyan in London where he worked for the Labour Party helping with the victory in the London County Council elections. He apparently met Harrisson by also trying to extract funding from Gollancz for a sociological project. Hood stayed at Davenport Street throughout the first phase of the Worktown project, leaving on 30 August 1938 by the wrong train (see Barefoot 1939: 201). He was later very positive about M-O, saying that it gave him a 'better understanding of human beings' and, more precisely, what 'made the working class tick'. Indeed, he went so far as to imply that the general postwar understanding of the working class 'came out of Mass-Observation' (Hood 1972: 6–8).

Eric Bennet was a local man who worked with M-O in Bolton. He was unemployed as a result of his activity with the Shop Assistant's Union and worked on unions and local politics for M-O (Barefoot 1939: 71). Other local recruits included Leslie Taylor, a Chemist's Assistant, who went on to be a full-timer with M-O in London and Bill

Rigby, an ex-miner. Another Boltonian, Harry Gordon, subsequently pointed out the advantage of the locals to M-O: 'I acted as a guide ... I found that the accent of some of the observers put some people off. About that time, the Lancashire dialect was more pronounced than today and people like me born and raised in the town were able to clear the air' (Gordon 1980). Gordon came in contact with M-O because he had been attending meetings of the Anti-Fascist Committee at the Davenport Street home of the secretary of the local CP branch. This highlights the close political links which, rather than Oeser's concept of functional penetration by rent-collecting, enabled the M-O operation at 85 Davenport Street to become part of the local community. Local communists and political activists had no difficulty working with fellow communists such as the Spanish Civil War veteran Sommerfield, Madge and, later, Geoffrey Thomas; or with nationally-known Labour activists such as Hood. Indeed, local Labour campaigns were run from 85 Davenport Street and investigations into methods of enticing non-voters to vote were indirectly aimed at making them vote Labour. The whole situation was not at all one-sided but a mutual encounter which could be disconcerting to the more middle-class observers, as Barefoot admits with respect to one local helper:

> When I first met Bill Naughton in Bolton, on some Mass-Observation assignment, I seemed to be getting on well with him and having a pleasant chat; and it was a nasty jolt when he told me, before leaving, that it had been interesting to meet me, because one didn't often get the chance of talking to people of a different class. I felt rather like a guinea-pig: the Mass-Observer mass-observed (Barefoot 1979: 5).

A reversal had taken place in which the intellectual, normally conscious only of his being-for-self had been made aware of his being-for-others. Naughton might never have gone on to write *Alfie* without having originally met M-O through delivering coal in Davenport Street, but the lives of the middle-class observers were no less transformed, while the heterogeneous – like Hood, university-educated and working class – and apparently classless – like Sommerfield – came into their own. Therefore, it can be seen that just this brief consideration of those named as involved in the projected Worktown publications – which should also include the Austrian Gertrud Wagner, who wrote the chapter on all-in wrestling for *Britain* – and some of their local colleagues gives the lie to attempts to brand M-O in Bolton as an intrinsically middle-class operation.

This is not to deny, as Gordon observes, that less regular visitors to the project sometimes had to be minded. Cawson recounts looking after Humphrey Spender on one occasion: 'He'd taken photographs of people in a pub and he was rounded upon by the landlord and he said, "Well, I wasn't doing any harm" and the landlord said, "Well, there might be someone here with someone else's wife! What right have you to be photographing here?"' (Cawson 1980: 9). A much longer and more apologetic account of the same occasion is provided by Spender in the notes to the photograph on page 82 of *Worktown People* (126–7) but Cawson's attitude is refreshingly breezy: 'What I think happened really was ... that people are really delighted to be taken notice of and to be asked questions.' In fact, much of the criticism of M-O's operations in Bolton is actually based on Spender's photographs, which are of course much more easily accessible than the contents of the archive. Here, Spender's surreptitious technique of trying to take photographs unnoticed by the subject is taken as representing a sociological voyeurism '"endemic in the methodology of Mass-Observation and of qualitative sociology generally"' which 'has the effect of reducing the observed ... to "the status of objects"' (Bennett and Watson 2002: 195; Hey 1986: 41–2). More specific criticisms, such as that of Raphael Samuel, focus on the actual representations: 'In Humphrey Spender's 'Worktown' photographs ... Images of entrapment abound, nowhere more so than at the pub, where the drinkers are literally cornered ... A bird's-eye view of Blackpool beach has all the faces out of focus, as though even on holiday relaxation was a forbidden luxury' (Samuel 1996: 331). Leaving aside the fact that even a cursory glance through *Worktown People* finds plenty of exceptions to this notion of universal entrapment, it can be argued that the core of the objection is that these photographs represent a combination of closure and nostalgia. This criticism is actually accentuated by a reading that sees the photographs as representing a resistant plenitude carved out amidst the grime of northern working life. The notion of a resistant 'everyday' colludes with the closure of the past in the name of the modern, as the two combine to deny the possibility of social transformation. Against this, it has to be said that Spender was only in Bolton for a few weeks and that his work is not really representative of M-O but of the documentary movement to which they were acting in opposition.

More typical was M-O's analysis in *First Year's Work* of the mass appeal of football pools:

Up to forty thousand watch the Saturday football match at the Northtown Stadium. It is the largest focus of a mass of people, about a fifth of the town's population. But a much larger number of Northtowners are less directly but more intimately associated with the game that is played there, and other games all over the country. As a Northtowner put it: "Now if it was a case of which would you have, Northtown to win and me to lose, or Northtown to lose and me to win, I'd say me to win of course." The "me" in this case, as in many millions of others, means his football Pool coupon (32).

Knowledge of the pools was found to be as essential as smoking and swearing for observers to be accepted when working in factories. It was both a form of social discourse and an acceptable expression of dreaming of a better way of life. While the odds of winning the big prize were apparently fourteen million to one – an uncannily familiar figure to the inhabitants of the present-day Britain of the National Lottery – it was 'THE ONLY CHANCE for the ordinary working man of ever getting a better time and independence' (37). The onset of the pools had changed everyday life in the non-conformist North 'in striking contrast to the slow advance or even relapse of other outlet-faith-hope-change of conditions systems such as the Church and the Labour Movement' (40). Interestingly, an analysis of Littlewoods, describing how people are 'brought in and kept in', clearly retains traces of the theory of the three social areas. While "Inside Information" from the 'gnome-like red "Little Ol' Man o' the Wood"' [Area 1] is supplemented by pictures of neighbours at the pillar-box posting their coupons [Area 2], 'from Area 3 come Lord Strabolgi, Nervo and Knox, and King John signing Magna Carta to lend their authority or popularity as an added inducement to join the Pools' (40). There is clearly some unease expressed here at individual desires supplanting local ties and creating a formless mass, which is then open to external leadership in dubious forms such as Littlewoods' 'The Chief': 'Very few people ever change from Littlewood's to other Pools' (41). However, this is outweighed by the recognition that the only other outlet for the desire of every working-class Worktowner to get out of the rut was the annual week's holiday. Therefore M-O conclude:

> The gigantic system of Football Pools, with its forty million turnover, is now something that cannot just be attacked. It exists. Simply to make Pools illegal would be a meaningless, negative gesture, which would have social consequences of the most unexpected kind – unexpected, because the legislators and leaders have no knowledge of how these things work, and what they mean in

terms of the people whose letters never get into the press, who often do not even vote for anyone. They are nevertheless human and necessary to be understood. We venture to think that towards that understanding this sort of research – and we have only been able to outline samples of what we are doing in Northtown – has something to contribute (45).

One such practical contribution took the form of showing a pools coupon to the Labour Party Leader Clement Attlee, who was in Bolton on a speaking engagement – photographed by Spender (H. Spender 1995: 89) – and had apparently never seen one before.

Concerned about the negative effect of the Pools on the Labour Movement – as an alternative hope of social change – M-O had printed an election leaflet 'designed exactly on the lines of a Littlewood's coupon' (45; reproduced in Harrisson 1961: 79–82). However, this concern with non-voters was to reach a new level of intensity with M-O's investigation of Jazz, in which they found dance hall attenders to be 12% more unlikely to vote than their peers. The whole history of dancing and jazz in relation to politics was the subject of Harrisson's Autumn 1938 *New Writing* piece 'Whistle While You Work', which analyses Jazz symbolism – 'continuously parallel with that of Auden-esque poetry, equally limited' (60) – and discovers it to be dream-magic 'nothing to do with the facts of everyday Worktown life' (62). Here again, however, rather than attacking the masses for being susceptible to such forms, Harrisson is primarily concerned with criticising the failure of political organisations to adapt to the Jazz-consciousness of the current generation. Although the Labour Party, the Clarion Cycling Group and the Left Book Club have all held dances, we are told: 'Political groups in Worktown regard the dance as a dance and a money source, show not the slightest appreciation of its propaganda potentialities, make no attempt to introduce any trace of politics into it' (58). His overall point is that anthropological knowledge highlights the orientation of Jazz to the symbolic order: 'Jazz has become or is becoming the religious ritual of post-War youth, and these songs of hope and happiness in a dreamworld every moon-night are the hymns of young England. This is our mass-poetry and a new folk-lore' (66). The implication of this was that if the political Left did not realise the significance of this situation, then the political Right would surely take advantage of it. An increase in the urgency of this situation was registered in *Britain*, in which Harrisson and Madge labelled both the Pools and Jazz as potentially Fascist forms.

First Year's Reception

May the Twelfth was published on 23 September 1937 at a price, 12/6, which took it beyond the pocket of most observers and consequently it only 'sold a bare 800 copies' (Calder and Sheridan 1984: 62). Jeffery states: 'Most critics have agreed that the book was something of a failure ... Papers of the right found a leftist bias in M-O, while left-wing journals were generally extremely hostile.' (Jeffery 1978: 24). However, this is somewhat inaccurate because there were favourable reviews in *Life and Letters*, *New English Weekly*, *Left Review* and *The Listener*, demonstrating that the world of the literary left journals from which M-O sprang remained a sympathetic constituency (see Various 1937: 166–8; Large 1937: 231–2; Richardson 1937: 625–6; Plomer 1937: xvi). Indeed, the false impression of a generally hostile reception rests on just three accounts: the reviews by G.W. Stonier in the *New Statesman* and Marie Jahoda in *The Sociological Review*, and the comments of T.H.H. Marshall in *The Highway*.

Stonier worked himself up almost to the point of hysteria in the face of a perceived insult to the individuality of the artist:

> That in literature facts do not of themselves add up does not trouble these writers at all. They have a mystical belief in the mob origin of their ideas, in mob infallibility, an absurdity to which their forerunners, the Unanimists, never descended. The mob in Mass-Observation, like the unconscious in surrealism, is always right; and as *faux naifs*, I should say Mass-Observationists have surrealism beaten every time. Naturally an art form from which no one is disqualified except the possessor of individual talent will appeal to many would-be writers (Stonier 1937: 534).

This was not an isolated attack and may well have been motivated by a knowledge of both, or either of, Harrisson and Madge's past criticisms of the Auden Group. Tension on this front continued to grow as Madge took advantage of the special 'Auden Double Number' of *New Verse* to complain that 'the original, but gawky, style has not developed' and commented: 'For myself, I would like to ask him to help with a Mass-Observation survey of Birmingham, his native city' (Madge 1937d: 27–8). This special issue then became one of the subjects, along with Auden and MacNiece's *Letters from Iceland*, of a review article by Harrisson in *Light and Dark*, an Oxford undergraduate magazine edited by Woodrow Wyatt – repayment, suggests Cunningham, for M-O work in Bolton (Cunningham 1988: 334–5). Harrisson notes that '[Madge] has already done too much to upset the Stoniers and the Audens by using M-O to write Mass Poetry, a horrible perversion and one which is now a part of the unforgettable past. Madge and myself now work on a common programme and are no longer concerned with literature – he got rid of that in the Coronation book' (10). His argument picks up from where *Letter to Oxford* left off by seizing on the timelessness of Auden and Spender's poetry, in order to argue that these ideas originated from the anthropologist John Layard, who Auden had known in Berlin in the late 1920s and Harrisson had worked with in 1936 following his return from Malekula (see Heimann 1998: 116–19). Harrisson suggests that these ideas 'are evidently tied up in an extreme regression from the liberal "idea of progress" into a long term cyclic stress which is also essential in fascism' (13). He then turns to a *News Chronicle* interview, in which Spender had said that he was intending to go to Bolton in order to see if the M-O material was suitable for writing a play from. The interviewer had asked, 'Is real life dramatic?' to which the reported reply was 'Yes, if the selection is done by an artist'. Harrisson comments caustically:

Apparently then, our job is to collect stuff so that the poets can pick the carrion out of it. All I can say is, Ha! Ha! Not all the dramas and dunghills on earth can affect the fact: that 'real life' is not what these boys mean these days and that 'dramatic incident', whether of poem, news or stage, is an 'isolate' from its social contacts of mass continuity (15).

This bitter, trenchant criticism provoked an exchange of views in the *New Statesman* that extended over four consecutive weeks. Firstly, Stonier retaliated with an article, 'Mass-Observation and Literature', which begins with speculations on the passing nature of fads from Rotariarians to nudists, before progressing to the perceived paradox of M-O that although it 'hopes to cut out literature by means of its accurate reports on human nature', the fact is that 'when a Mass Observer gets to work, as Mr. Harrisson does on poetry, he is apparently quite incapable of being accurate' (Stonier 1938: 327). This brought Madge into the correspondence columns in order to good-naturedly defend himself against both Stonier and Harrisson:

Mr Stonier has not brought my published poems into the argument, but in them consists my right to interject on the question of whether Mass-Observation is trying to "cut out literature." Tom Harrisson tells me he cannot understand why I should bother to write poems at all; perhaps he regards them, mistakenly, as part of my "unforgettable past." I regard his letting-off steam in undergraduate papers as an amiable kind of nonsense about what is, after all, mainly a rather pretentious kind of nonsense. Far better, surely, than seriousness about nonsense such as anthologists of modern verse are prone to. Mr. Stonier's trouble is, I feel, that he will take the nonsense seriously and thus make nonsense of the essentially serious applications of Mass-Observation (Madge 1938a: 365).

The next week, Harrisson returned to the fray by arguing that he only wanted to supersede literature's right to instruct, and not its right to communicate experience, 'but I suspect that Stonier will fight to the last ditch, slinging mud all the way, to preserve his long-held prerogative of instruction' (Harrisson 1938b: 409). Finally, even Spender joined in to sarcastically inquire as to whether any number of 'statistics about fried fish shops' proved that poets writing about death in an age of mass war were unrealistic? He took a moral line:

About a year ago today I stood in a field near the village of Morata near Madrid, where there were a considerable number of the dead. They lay quite still; they did not eat fish and chips, discuss betting pools, back horses, look at girls crossing roads, tap the ends of cigarettes before lighting them, whistle in public lavatories, or do any of the things that Mass Observers write down in their notebooks. I doubt whether Messrs. Tom Harrisson and Charles Madge, if they had been there, would have noticed them, since, in their passion for the explicit, there seem to be several implicit phenomena which they do not remark. If they had noticed them, they would have had nothing to report (S. Spender 1938: 477).

One apparent upshot of this exchange was an attempt by Madge to get Spender to arrange for him to go to Spain, as suggested by the postscript of a letter from Madge to Spender's wife, Inez, later that year: 'Has Stephen done anything about sending me to Spain? Could we send a Mass-Observation unit, a small one, to observe an air raid?' (Madge 1938c). This request was complicated by the circumstance that Madge and Inez had become lovers in the meantime. They would eventually undergo respective divorces and marry each other. However, even in the context of the open relationships enjoyed by bohemian intellectual couples like the Madges and the Spenders, there was already sufficient friction by October 1938 for Spender to brand Madge a 'crook' who seduced women with 'Wykehamist flattery' (cited in Sutherland 2004: 252, 582). The short term consequence for M-O was that Madge switched houses with Harrisson and took up residence at 85 Davenport Street from the beginning of November. The long term consequence, as previously discussed, has been the continued misrepresentation of M-O and its founders in literary accounts of the 'Thirties' such as the recent biography of Spender, which inaccurately describes Harrisson as an 'Oxford-trained anthropologist' and maliciously accuses Madge of 'undeviating Stalinist orthodoxy' (Sutherland 2004: 239–40).

Sociological reception has been equally damning ever since Marshall, Reader in Sociology at the University of London, set the pattern with his brutal dismissal of *May the Twelfth*: 'A four hundred page volume has been published which is so completely devoid of interest that even the most well-intentioned reviewer is at a loss to find anything to say about its contents' (Marshall 1937: 48). He followed this up with an attack on the fundamental M-O principle of using untrained observers: 'The patient with an amateur knowledge of medicine is more likely to hinder than to help the doctor in his diagnosis' (50). The review by Jahoda – one of the authors of the 1933 *Marienthal* study – replicated this twin attack:

> They seem almost to think that exhaustive observation and description are possible. This is nonsense; although the book is full of details, it cannot compete with abundant facts and details of reality... It is true that to abstract, one must know the complex reality with all its details. But no representation of details dispenses with the necessity for abstraction .... I have little doubt that every observer was enthusiastic about his task and wished to be quite objective and unbiased. As, however, they are all casual and untrained observers, they must show bias in their observations (Jahoda 1938: 208–9).

However, the vehemence of these positions does not reflect the strength of an established discipline so much as a discipline struggling to establish itself. Dennis Chapman later suggested that Madge and Harrisson's success in gaining the support of a wide range of scientific intellectuals such as the zoologists Julian Huxley – who wrote the foreword to *Mass-Observation* – and James Fisher; the social anthropologist

Bronislaw Malinowski; and the economists John Jewkes, Percy Ford and Phillip Sargant Florence; was experienced as a threat by British sociology at a time when there was only one university chair in the subject (Chapman 1979a: 6).

According to Chapman, a volume being prepared by a working group of sociologists, anthropologists and psychologists, who had been meeting since 1935, to formalise methodology and approaches to social science, turned into an academic reply to M-O. The preface to *The Study of Society* commences with arguments, which by now must have been familiar to an educated readership as the mantra of the times:

> There is today widespread recognition of the fact that the future of human civilisation depends to a high degree upon Man's capacity to understand the forces and factors which control his own behaviour. Such understanding must be achieved, not only as regards individual conduct, but equally as regards the mass phenomena resulting from group contacts, which are becoming increasingly intimate and influential (Bartlett et al 1939: vii).

The second sentence is clearly a reference to the impact of M-O. As is the subsequent insistence on the importance of specialised training, which is followed by the sentence: 'At the same time it is recognised that invaluable services can be rendered by persons who have had little opportunity to obtain technical instruction, and at various points in the book attempts have been made to indicate how the amateur investigator can best assist in the development of social studies' (viii). For 'amateur investigator' read mass-observer. It becomes rapidly clear that a collective decision must have been taken by – or imposed on – the working group not to mention M-O by name, despite the fact that it included Blackburn and Oeser who had actually worked with them. For instance, in his chapter, Oeser refers to the number of occupations held by Harrisson in Bolton in order to illustrate that 'single investigators may take various jobs in turn' and then in comparison points out how much better a team such as his in Dundee could functionally penetrate society (Oeser 1939: 411). As he knew perfectly well that Harrisson had gone on to form a team which was proving remarkably successful not just in penetrating society but in establishing an unprecedented mutual linkage with it, it must be assumed that the omission is deliberate. Of course, it might have been that M-O did not sufficiently represent Oeser's preferred model of team organisation, which was to set it up to mirror the hierarchical structure of the society to be penetrated so that, for example, 'the team's executive and administrative head ... can enter into relations with the heads of government departments and other officials' while the team workers could mix with the workers in the community (413).

However, this state of denial reaches its most absurd depths in the chapter 'Suggestions for Research in Social Psychology' by F.C. Bartlett, one of the editors and the Cambridge Professor of Experimental Psychology: 'I shall first consider problems for the psychologist who is working in the laboratory; then problems for the psychologist in the field, and finally problems towards the further definition of which persons who have not necessarily received any specific training in psychology may nevertheless make important contributions' (Bartlett 1939: 24–5). After fifteen pages on laboratory work, Bartlett manages two and a half on field work by means of discussing the 'topical' question of 'Problems of Leisure' – obviously chosen because M-O were investigating it – before concluding: 'In some cases relevant questions have already been made a topic of laboratory investigation, and part of the technique for their initial field study is therefore at hand' (41). The role of the amateur investigator in this brave new world of social research was to undertake the preliminary collection and classification of evidence necessary for the experts to formulate social problems in such a manner that they might be 'successfully attacked'. Areas where the co-operation of the amateur investigator would be welcomed might include studies of a particular district focusing on entertainments, social co-operation and political attitudes. In particular, 'something could be done by a systematic analysis of leading articles in popular newspapers especially when, as in times of crisis, nearly all newspapers are dealing with the same situations' (43). The chapter concludes with the recommendation that central 'clearing-houses' be established, allowing the amateur investigators to be directed and kept informed of results from other areas.

What is clear from this farcical nonsense is that the history of British sociology could have been very different with a slight increase in the courage and imagination of its elders at this point. There were obviously voices on the working group arguing for the formal incorporation of M-O into an integrated approach to social sciences. For example, T.H. Pear, Professor of Psychology at Manchester, provides a coded endorsement of M-O in his own chapter, arguing that the varied uses of the coronation for propaganda would make a useful social investigation, before going on to actually break the embargo by suggesting that 'it is unlikely that the cinema will escape "Mass-Observation" (Pear 1939: 2, 15). However, rather than make the collective leap to endorsing M-O as the obvious complement to academic social research, the working group and the editors ended up taking refuge in this ridiculous compromise whereby the ideas behind M-O were discussed in such an abstract, hypothetical and condescending manner as to be of no use to anyone.

No such timidity was shown by Malinowski, who eagerly seized on Madge's offer to write the afterword for *First Year's Work*. The departure of Jennings had altered the balance of M-O and shifted it to the Madge-Harrisson scientific axis as outlined in the second section of *Mass-Observation*. Within less than two months of publication, the Bulletin for November was characterising the 'May 12th book' as representing 'a stage of M-O that has now been left behind' (M-O A: FR A4). As previously discussed, Madge had begun to cultivate a scientific public persona by attending Malinowski's seminar and it is clear, following the commercial failure and mixed critical reception of *May the Twelfth*, that gaining Malinowski's endorsement was important for restoring M-O's credibility. In turn, Malinowski gained the right to put down his rival Harrisson – still enjoying public recognition as an anthropologist from *Savage Civilisation* – and take some of the credit he felt he was owed for the concept of anthropology at home (see MacClancy 2001: 91–5). Therefore, the afterword was taken seriously on both sides and came to dictate the publication date of *First Year's Work*, as it grew to 16,000 words in length, completed with Madge's help in 'long sessions' at Malinowski's home (Madge 1987: 76).

'A Nation-Wide Intelligence Service' can be read not only as a statement of Malinowski's position with respect to M-O but also as a public act of self-fashioning on Madge's behalf. The tone is set in the opening paragraphs, where Malinowski praises the 'movement' for 'humility ... in its readiness to co-operate with other scientific workers' and 'its ability to reconsider its aims and to reorganise its collective methods of research' (Malinowski 1938: 83). It quickly transpires that M-O's former heretical deviations, confusing the relationship between subjectivity and objectivity and thus being insufficiently scientific, are deviations from the true path laid down by Malinowski himself: 'I feel that in a way I have been responsible to a large extent for the inevitable consequences in the development of the functional method of anthropology: I mean, for its definite move towards *Anthropology Begins at Home*' (103). He sees M-O, once corrected in its errors, as having the potential to realise his own dream of ethnography determining 'a correct theory of society for the future scientific guidance of human affairs' (104). These necessary corrections are that observers should be properly regarded as informants in the ethnographical sense (96, 118) and that M-O's investigations, questions and instructions to observers should always be oriented by function (105–7, 111–12).

The change in meaning entailed by the switch from the original conception of the observer to that of informant is subtle; analogous to the slippage between participatory and representative democracy. Malinowski's position is that while the more subjective it is the better, the subjective data of the informants has to be subject to objective – disinterested – analysis by somebody outside the grouping in question; as opposed to M-O's hitherto position of 'confusion which would require the observer and the observed to be one and the same person' (92–8, 99). However, this was precisely the most radical element within M-O and the necessary condition for the possibility of collective independence generated by its coronation pastoral. Accepting Malinowski's criticism would limit this independence to individual ethnographers above and beyond the society whose observations they studied.

The major faultline that subsequently developed between Madge and Harrisson was due to the fact that Madge was prepared – at least outwardly – to accept this criticism and take the pathway of a respected social scientist while Harrisson was not. One way of assessing what is at stake here is to look at Malinowski's criticism of Harrisson's claims, in *Mass-Observation*, that in order for an anthropologist to understand other people he has to 'learn to dance their dances, eat their diet [and] love their women – factors at least as important as language' (Madge and Harrisson 1937: 35–6).

> I ... understand from my native informants, male and female, that when a white man 'makes love' to a native woman, his is not an ethnographic experience. The white man cannot transform himself into a Melanesian. There are either no physiological differences in some of the bodily functions, in which case there is no need to duplicate them, or, if there are racial differences, the white man, as an observer, has to draw the line (Malinowski 1938: 99–100).

There is only a line to be drawn if 'the white man' regards himself as essentially different – no less in his 'whiteness' and 'masculinity' than in his scientific status as ethnographer – from 'the native woman'. Such a difference can only be maintained by rigorously excluding the 'other' but it takes refuge behind moral hypocrisy. John Roberts has described this, in a related context, as a process of 'middle-class intellectual production' generating 'an actual *fear* of theoretical engagement with everyday culture' (J. Roberts 1998: 70). A similar process can be seen in those contemporary criticisms of M-O for reducing the status of the observed to that of objects, which tell us more about the fears of the critics – of being made conscious of their own being-for-others – than about M-O.

These differences meant nothing to Harrisson and, therefore, he drew no lines. Moreover, as shown in *Savage Civilisation*, he was well aware that ethnographers had no problem in crossing the line in private when it came to writing from the 'native' point of view and that they only denied it in public in order to assert their individual professional status. By sending up this duplicity with accounts of cannibal feasts and sex, he laid bare the power of writing to transcend difference and sought an identity that was not dependent on bourgeois individualism. Madge later suggested that Harrisson took this approach from Orwell despite his tendency to run down *Wigan Pier* in public. This is a shrewd analogy because it recognises how a shared reliance on textual performativity underwrites what is essentially the same claim: that a former public-school boy can meet and understand people from the working class on equal terms. However, Madge's description of Harrisson's professed attitude – 'that one had to be completely ordinary and that our whole aim was to identify ourselves with people as ordinary as possible' – as 'frightful' is also telling about himself (Madge 1978a: 2). What is certain is that they both took different attitudes to crossing the line in public and consequently the idea of M-O as observation of the masses by the masses for the masses began to unravel.

The Munich Crisis

M-O's newfound scientific respectability had several discernible consequences. In July 1937 they had announced: 'Early in 1938 a conference will be held to which all Observers who have worked for over six months will be asked to come. A series of local conferences has already begun' (M-O 1937a: 5). Eight months later, in his intervention in the Stonier debate, Madge wrote: 'I want to make it clear that Mass-Observation is not an organised faction. Unlike Buchmanites, Rotarians, Wandervogel and Nudists, Mass-Observers do not hold meetings: in the interests of science they are discouraged from doing so' (Madge 1938a: 365). The national conference for Mass-Observers had never taken place. All that remained of the PP program was a big red question mark pencilled into the margin of the abandoned draft of *First Year's Work*. The published version contains no trace of any commitment to 'educate [the masses] into awareness of their own potentialities and so render them less vulnerable to anti-social propaganda' (M-O 1937b: 3–4).

The theoretical framework of the three social areas began to undergo a rapid transformation. Although never mentioned again after the publication of *First Year's Work*, their trace can still be detected in M-O's later work and even in Madge's post-war writing. By looking at the revisions made to the earlier draft of *First Year's Work*, it is possible to see this transformation taking place. Common to both draft and book is the sentence: 'Already an empirical knowledge of the factors involved in the changing of social habits is being acquired by advertisers and political propagandists' (M-O 1937b: 3; Madge and Harrisson 1938: 9). However, the draft's subsequent assertion 'that man has a natural tendency to conform, by agreement or by imitation with other members of his social group', is converted by the published version into an assumption held only by political propagandists and advertisers (4, 9). This means that the concept of conformity changes from being 'a question of a relationship in terms of area 3' into 'an intervention from area 3' (7, 11). Area 3 has changed from being part of the social relations of any individual's life into an external force, which imposes itself on the individual. This has two consequences. The first is that these external forces of conformity are now represented as something that can be studied completely empirically: 'The social factors become "invisible", until investigation brings them out' (Madge and Harrisson 1938: 23). Secondly, agency is being denied as the individual is portrayed as passive before the external intervention of advertising and propaganda. In their study of opinion formation and social habits, M-O no longer recognised the potential of a conscious mass society capable of resisting external pressures to conform and thus they were forced to look for other means of protecting the masses from anti-social propaganda.

In the Penguin Special, *Britain by Mass-Observation*, published in January 1939, this search is situated in the framework of a structural analysis: 'What happens in the political sphere obviously affects the sphere of home and work; equally obviously, political developments are affected by the reactions of ordinary people. But between the two there is a gulf – of understanding, of information and of interest. This gulf is the biggest problem of our hugely organised civilisation (Madge and Harrisson 1939: 25). It can be seen how the understanding of the media's role in determining social relations, which had originally been gained by analysing the Coronation in terms of the three social areas, had evolved since *First Year's Work*. Study of the relationship between the king and the people had made visible the ability of the media to bridge the gap between areas 1 and 3 and thereby erode the role of area 2 – a process which Jennings had contested by deploying himself in the image space and showing this evacuated public sphere as still a site of contention. However, in *Britain*, Madge and Harrisson accept this erosion of area 2 as a *fait accompli*: there is only a 'gulf'. On one side is the everyday world of 'home' and on the other side are the opposing forces of capitalist everydayness – advertising and political propaganda – waiting to come down like the wolf on the fold. Consequently, the impression given to the reader is not just that a decrease in the contestation of everyday life has been registered, but also that any sense of everyday life as a contested category has been lost.

This urgency comes from the recognition that Western society has already found one solution to the divide between the leaders and the led: 'Hitler, in a speech at Nuremberg, pointed to one possible answer when he said to 180,000 followers: "You can rely blindly on me, just as I can rely blindly on you [....] I am the spokesman of the German nation and I know that every one of a people of millions agrees with my words and confirms my view"' (25). This analysis anticipates Sebastian Haffner's argument that the only way to understand the phenomenon of Hitler was to consider 'German and European history as part of Hitler's private life' (Haffner 1940: 16). The broadcasts of the Nuremberg rallies were a means of replicating and perpetuating the moment of the coronation in which areas 1 and 3 had been brought together in mutual affirmation. The German population had become incorporated into Hitler's area 1, as M-O might once have recorded the situation. Therefore, Harrisson and Madge's identification of a similar gulf between politicians and home in Britain caused them deep concern, which was intensified by the apparent evidence of a decreasing public interest in each successively occurring national crisis: 'with the crisis, a flock of other abstractions appear in the newspapers to perplex the public mind – "tension" becomes greater or less, the "situation" gets better or worse, the "problem" is solved or unsolved. To most people these abstractions are as mystical as the voice of the stars' (25–6). However, their problem was – as outlined above – that they could no longer identify a public sphere in which to contest or mediate this depoliticisation of the masses in their own homes. Therefore, all they had to go on was what they had learnt about opinion formation in relation to advertising and smoking: 'Opinion is made in two ways. It is made by each single person looking at the facts, as far as they are available, and then framing his own judgement on them. It is also made by the reaction of each single person to the opinions of other people' (32). The resulting anti-fascist strategy called for objective facts to be made readily available to everyone while simultaneously demanding that the media should accurately reflect public opinion formed on the basis of these facts. Under the perceived circumstances, it was an ingenious solution to the problem of the gulf between the leaders and the led, reversing the current of the feedback loop of fascist affirmation by preventing leaders from including the masses within their own sphere of private opinion and making them answerable to public opinion instead.

Madge and Harrisson illustrate their concern with the role of the media in opinion formation by dissecting the press coverage of the three meetings between Chamberlain and Hitler, when public opinion swung wildly without anyone's knowing what had been discussed or agreed. Chamberlain's first flight to Munich on 15 September 1938 is characterised in mythic terms:

> Everyone was fed up and longing for the thing to stop when there came the wonderful news that our Prime Minister, Chamberlain, was going to fly to see Hitler. It was emphasised that he was 69 and that he had never flown before. The combination of his age and his sky-journey made him a father-deity. But his self-sacrificial abasement in going to Hitler made him like the son-deity who descends among the wicked to save them. The piled-up suspense and anxiety could only be dispelled by a gesture of this super-human kind, by a piece of myth-making. The whole Press applauded ... (63–4).

As they note, nobody knew what was talked about but 'a sensational swing of opinion in favour of Chamberlain resulted' (64). However, they are able to show from continuing M-O research throughout the crisis that this was based on the assumption that Chamberlain was putting his foot down: ' *Man of* 30 ... Either it comes off or else we know it's got to be war. Chamberlain will have told him that now' (67). In fact, according to A.J.P. Taylor, Chamberlain afterwards admitted, 'In principle I had nothing to say against the separation of the Sudetan Germans from the rest of Czechoslovakia, provided that the practical difficulties could be overcome' (Taylor 1964: 217) and he had flown to Munich alone just in case the French proved intransigent on the point. By the 22nd and 23rd, it was clear that this had filtered through to the public and popular opinion was starting to run ahead of the newspapers, which had been 'soft-pedalling' since the 19th: 'For the moment, until we have full official confirmation of the nature of the proposals, it is better, after making that fact clear to reserve comment ... (*Daily Herald*)' (71–2). On the 21st and 22nd, M-O had asked 356 people what they thought. The tenor was set by a bus conductor in Lewisham: 'What the hell's he got the right to go over there and do a dirty trick like that' (77). Madge and Harrisson derive statistics from those two days that show percentage increases of those indignant from 36% to 44% and of 'don't knows' from 25% to 32%; and percentage decreases of those who were pro-Chamberlain from 25% to 18% and those who thought there would be no war from 14% to 6%. This leads to their scathing attack on the Press:

> During one whole week, no outsider reading an English newspaper could have guessed that an increasing proportion of the population were feeling once more increasingly bewildered, fearful and ashamed. The readers themselves didn't guess it in many cases ... The fact that the papers hung back has a delaying effect on public opinion, because newspapers are so much looked to for social and talk sanction ... By representing pro-Chamberlain as the universally felt sentiment, (when in fact even at its top point he never scored more than 54%), individuals in their homes were temporarily made to feel that being anti-Chamberlain was old, anti-social, or Socialist – until at work and in the streets, by the third day each had gradually found hundreds of others agreeing in this secret shame (105–6).

This information could have been used to demonstrate how collective activity in area 2 – public contact in the streets and workplaces – can be a source of resistance against the official values being circulated in area 3. M-O, by abandoning that framework, were no longer able to make this analysis despite recording what was happening in practice. Instead, far from emphasising this nascent political activity in order to encourage it, they treat it as evidence that the main result of the papers' 'hanging back' was the creation of a vacuum that allowed fears to enter the public realm.

Rather than the press reflecting the public sense of indignity and shame at the fate of the Czechs, the public was bombarded with details of war preparations: 'On September 27 the mechanism of ARP began its turgid rotations .... It was at this stage that mass-fears began. As one observer put it, "with distribution of gas masks began real fear..."' (87–8). On that evening came Chamberlain's speech about how horrible it was to be digging trenches and how 'inconceivable that they should undertake a war because of a quarrel in a far-away country between people "of whom we know nothing"' (91). Against this mass-fear and tension, Chamberlain's third flight appeared as a miraculous *deus ex machina* – bringing scenes of hysteria to the House of Commons and a huge surge in public support for 'peace in our time'. Chamberlain had replaced 'the Golden Bough by a paper symbol' (102). Harrisson and Madge assert that this mythic resolution was a product of the press:

> Thus at this climax in national history, and although the time of the Premier's arrival had been announced over the wireless, there turned out under half the number of people that can be counted on for a routine Communist rally in Trafalgar Square (and there are 16,000 members of the Communist Party).
>
> No second division football club could survive on a Chamberlain gate. Nevertheless next morning the Press arranged photos and headlines which gave the impression of enormous crowds; with captions typified by the *Mirror's*: 'The enthusiasm of the nation was led by the King ...' (103).

They go on to note that 'all through the crisis fantasy and fact become entangled' (113). The suggestion is that the threat of war, while real enough, had been experienced at a fantasy level and then dissipated with a fantasy resolution which allowed what had been objectively unacceptable (the forced partition of Czechoslovakia) to become publicly acceptable. While this analysis of the effects of the press representation was accurate and penetrating, it still left them nothing to advocate other than better press representation.

Furthermore, it is not clear how such better representation could be possible in terms of their own analysis because, in practice, this was only enabled by the subjective component in their reporting of public experience. In other words, it is the blocks of reported conversation from people in the street that substantiate their arguments. Yet, they were calling for the reporting of objective opinion in a quantifiable manner acceptable to social science. Their inconsistency on this point was highlighted by the woeful inadequacy of their own statistical analysis. They had neither used random sampling techniques nor interviewed the same people over time in their opinion surveys – faults that were seized on by contemporary critics (e.g. Firth 1939: 182–6). Calder commented 'It is never wholly clear, from *Britain* or from other M-O publications, whether the primary aim was observation *of* the masses or *by* the masses' (Calder 1986: xiii). However, in this case the ambiguity simply reflects the confusion between publishing opinions tendered by the masses and tabulating mass opinion as a subject for quantitative treatment. In attempting to comply with the demands of Malinowski and their critics, M-O were actually producing the confused results that had supposedly been inevitable from their original methods. These contradictions crucially affect the ultimate conclusions made:

> Through all our research results the interest in oneself and one's own home has predominated far and away, over international and general political concerns, except in the upper middle class. Until the last week in September, 1938, the great mass of working people were never observably affected in any marked degree by any crisis within the eighteen months during which we have been making observations, and it was only when the international situation threatened to enter their own homes, as gas, that a real mass response was apparent. In this world of rearmament and fortifications, an Englishman's home is still his castle, although well under half are their own landlords (217).

From a chapter entitled 'Castles in the Air?', this passage demonstrates how Madge and Harrisson have begun to represent the everyday resistance of 'home' as an obstacle. The biggest problem is no longer perceived as the 'gulf' between home and politics, but the tendency of home to shut out the world and isolate itself. At the beginning of the book the question is posed whether Britain was really 'tensely watching' the crisis over Czechoslovakia or the racing news and the daily horoscope (7). By the end, the implication is that such activities are the practical shapes in which Fascism invades the politically isolated home:

> New party alignments are indeed the subject of some intrigue at this moment. But other new parties have escaped the attention of politicians and intellectuals. Cads' College, Narkover, Littlewoods' Loyalists ....
>
> .... At the head of most of these organisations is a dictator, Clifford Whitely, Will Hay, Gracie Fields, the Western Brothers. Football pools specialise in this mechanism, the two biggest having respectively The Chief (Cecil Moores), and The Governor (Vernon Sangster) .... With these modern techniques and in a language distinctly different from that of the House of Commons, new and

potentially powerful groupings are being formed, almost unnoted, throughout the structure of English society ... (243–4).

Of course, rejecting the everyday as a basis of political resistance is consistent with M-O's founding position. Originally, it was the feature that had differentiated them from the documentary movement. Moreover, it underwrites an acute political perception unique to Britain at that time, having more in common with continental figures such as Breton and Benjamin. Alone in Britain, they were able to backup criticism of Chamberlain with a devastating demonstration of exactly how the National Government's resolution of the Munich Crisis had been, in effect, a fascist resolution: 'In the atmosphere which the above organisations imply and which was vividly accentuated by Chamberlain's flights, it was for a few days possible for many to look on the Prime Minister as a new leader to bind together the disintegrating or novel loyalties' (245). Yet the problem was that they were struggling to find any real alternative to such fascist processes, other than utilising similar 'modern techniques' to the 'new parties' in order to penetrate the politically isolated home before the real fascists: 'Only an extremely vigorous and immediate intervention along new channels and with new techniques, by "democrats," "scientists," etc, can perhaps provide an alternative to this impending dictation' (245).

The Lambeth Walk

Britain provides one positive approach to popular culture: 'As a symptom of changing social attitudes, the Lambeth Walk points the other way from Football Pools and Daily Horoscope' (174). This is because, Madge and Harrisson argue, it is a trend originating in the working class and not one foisted upon it. Therefore, it has potential as a model for the new type of intervention they are looking for: 'We may learn something about the future of democracy if we take a closer look at the Lambeth Walk' (140). This argument is backed up by one of their longest and most thorough analyses. Early on the chapter is an account of the 'native cockney culture ... still vigorously existing' (143) embodied in a darts club attached to a pub, which organises parties after closing time on Sunday nights in members' houses:

Obs.: 'What are some of the other dances called?'
A.: 'We've never known the names. We just do'em ourselves. For instance, the chap who's just gone out to sign on, he wants to be a dog. If he's half drunk, he wants to be a dog. He wants to bark....

Then we has bloomers and blouses, we dresses up in them.... The other night we had the women on the floor, fighting over her like two dogs. We don't do it very legal you know. I come home with a black eye' (144).

Harrisson and Madge are interested in how the Lambeth Walk has spread from such origins to 'Mayfair ball-rooms, suburban dance-halls ... Scotland ... New York ... Paris ... Prague' (139). They identify the immediate source of the popularity of the craze as the December 1937 show, *Me and My Girl*. This involved the cockney comedian, Lupino Lane, playing a Lambethian who 'inherits an earldom but cannot unlearn his cockney ways. At a grand dinner party he starts "doin' the Lambeth Walk" with such effect that duchesses and all join in with him and his Lambeth pals' (140). Empson's analysis of pastoral is invoked to point out that the show is essentially about 'the contrast between the *natural* behaviour of the Lambethians and the affectation of the upper class' (157).

The show includes a scene mocking the Coronation with Lane in his peer robes, at one point lying on the floor as though to suggest a state funeral. This sounds like the perfect example of Empsonian pastoral with Lane at the third level of comic primness. It is not clear if he is satirising, accepting or simply innocent of the importance of traditions: he has complete freedom of action. However, the dance craze was the result of the managing director of the Locarno Dance Halls seeing Lane's typical cockney walk – 'a swagger and roll of the shoulders' – and getting his 'ace' dancing instructress to elaborate it into a dance (141). Tellingly, this became so popular that she then had to teach Lane how to do it (161). Madge and Harrisson provide a list of the stages that the Lambeth Walk has gone through, which can be summarised:

1) The native cockney culture and related cockney walk.
2) Lupino Lane and the show.
3) The Dance Hall manager who had the walk developed into a dance.
4) The BBC and press which gave the dance publicity.
5) The masses who took up the dance with enthusiasm.

They conclude:

Of these five factors, 2, 3 and 4 represent the Few who cater for the Many – in this case successfully. Factors 1 and 5 represent the influence of the Many. The cockney world of Lambeth – its humour, its singing and dancing, the way it walks – is a mass-product with a special local character. But this character is strong enough to appeal to a much wider mass of people as soon as it is made known on a wide scale (141–2).

This illustrates the critical inadequacy of concepts such as the 'Few' and the 'Many': clearly there is something amiss with an analysis which characterises the press and the BBC among the few. If the analysis had been made in terms of the three social areas it might have looked as follows (with analyses in terms of everyday life in parentheses):

1) Area 1 (the everyday.)
2) Areas 1 and 2 (the performance of the everyday in the public space of the theatre, thus contesting an homogenised everyday life and creating the possibility of independence.)
3) Area 2 (a business relationship between employer and employee creates a dance that functions in a formal social setting – the dance hall.)
4) Areas 3 and 1 (the media penetrates the everyday.)
5) Areas 1 and 3 (the dance is incorporated into an homogenised everyday life.)

The symmetry of the published analysis is shown to be false. The fact that the 'Many' add their own ways to the mass craze demonstrates the functioning of an undifferentiated everyday life rather than any organic link to native cockney culture. With the exception of its origins, the media-promoted dance craze is not structurally different from the other popular culture manifestations discussed in *Britain*. The extent of the departure from *May the Twelfth* can be seen from the judgement: '... the Lambeth Walk succeeded in a big way, because it makes everyone do the same thing at the same time, and express their togetherness with smack and shout' (183).

Some examples of dissent are provided: 'It leads to the viewing of slum-life with all its poverty, dirt and misery through the rosy spectacles of the wise-cracking Cockney and the glamorous Pearly King. It is a common dodge to make us laugh at our miseries and put them out of mind that way' (169). The footnote in which this complainant is answered is worth quoting in full:

He is objecting to the Lambeth Walk because it induces the feelings which Empson calls 'pastoral' between rich and poor. Empson says it would annoy a Communist to admit that W.W. Jacobs is a proletarian author: 'Probably no one would deny that he writes a version of pastoral. The truth that supports his formula is that such men as his characters keep their souls alive by ironical humour, a subtle mode of thought which among other things makes you willing to be ruled by your betters; and this makes the bourgeois feel safe in Wapping.' But as we shall see, the working class has taken up the Lambeth Walk with more enthusiasm than anybody – a fact recognised and made use of by both the Communist Party and the Labour party. In the latter case, it was partly due to a long discussion between a leader of M-O and the Transport House [Labour Party headquarters] propaganda experts, who could not see the faintest connection between the Lambeth Walk and politics until the whole history of dancing and jazz had been gone into (169n).

The implication is that people would be willing to be ruled by their betters provided they were allowed suitable modes of expression and outlets for personal enjoyment. Harrisson seems to have been some sixty-odd years ahead of his time, even inducing the 'propagandists of Transport House [to produce] a song (to tune of the Lambeth Walk) for use at Elections ...' (176). Is this really the possibility that illustrates the potential of the Lambeth Walk as a new 'democratic' intervention against potentially Fascist new party alignments? As one reviewer observed at the time:

> ... if the Mass-Observers continue their inquiry into the dominant factors in determining votes, each party may find itself obliged to modify its policy .... With increasing knowledge of the irrational responses of the electorate, each party is likely to become still more inconsistent in its programme; and if politicians also borrow from manufacturers their scientific methods of bamboozling the public with advertisements, democracy may become unworkable (Mortimer 1939: 62).

However, prescient though this undoubtedly is, the immediate political problem in 1939 was the rise and spread of fascism. The importance of the Lambeth Walk was that, even at the height of this critical juncture, it seemed to hold out the possibility of squaring the circle between the leaders and the led in Britain, as Highmore argues:

> If the section on the Munich crisis offers an example of the gap between representations of political events and responses to those

representations in everyday life, then the discussion of the Lambeth Walk demonstrates another reading of the politics of everyday life distinct from the cultural practices associated with Fascism. It is not a 'scientific' critique of Fascist culture; it too is bound up in ritual and superstition, and for this reason offers an alternative imaginary identification that can be seen as (effectively) resistant to Fascism (Highmore 2002a: 108–9).

Highmore's point is that the Lambeth Walk offers a rival democratic myth to Fascism. It is equally ritualistic but this ritual is orientated to a radically different symbolic order. Citing the example of the Lambeth Walk being used to break up a Mosleyite demonstration, Highmore suggests that it 'can be seen as the exemplification of a Popular Front culture that comes out of the everyday' (108). However, what distinguished it from the prevalent Popular Front emphasis on resistant everyday plenitude was that its essence as performance offered roles to other social classes in a way that other representations of the working class, such as strikes or hunger marches, could not however well documented or dramatised they might be. Madge and Harrisson were able to understand this because they experienced it for themselves, as Madge describes in a letter of the time: 'Then I went to Lambeth and Christ! What an evening. We had the most wonderful party till two in the morning. Tremendous dancing, tremendous people. All playing piano or accordion by ear. All handing round their glasses of beer for others to drink. Transvestism and fine class-conscious songs' (Madge 1938b). This suggests little inhibition in private about following Harrisson's prescriptions for the anthropological study of other people. In fact, Madge seems to have spent most of the first half of August 1938 attending different Lambeth Walk events – presumably he contributed much of the actual research in the chapter – and speculating as to whether the real attraction of the actual 'Walk' was that it combined the aggressive swagger of the hard man with the sway of a slinky woman. It even inspired him to attend CP meetings regularly again and to go out and enjoy selling the *Daily Worker* on the Honor Oak Housing Estate (Madge 1987: 82). Doing the Lambeth Walk combined the carnivalesque liberation of coronation pastoral with the Worktown experience of observers discovering their own being-for-others. This was because it was, as Madge and Harrisson correctly identified, a version of pastoral:

When you do the Lambeth Walk, you pretend to be a Lambethian .... One thing which the huge popularity of the Lambeth Walk indicates quite definitely is a very widespread 'wish to be these people', though of course that wish is not a simple or straightforward one, and includes elements of make-believe and ballyhoo. The upper classes wish to masquerade as Lambethians: sixteenth century lords and ladies played, in pastoral make-believe, as shepherds and shepherdesses. The middle classes wish to be Lambethians because it temporarily lets them off a sticky code of manners which they usually feel bound to keep up. The working classes wish to be Lambethians because Lambethians *are* like themselves, plus a reputation for racy wit and musical talent – partly they represent that part of the working class which knows how to have a good time (173–4).

It was this ability of the Lambeth Walk to 'unite without unifying' that allowed *Britain* to represent the nation as simultaneously socially heterogeneous and the site of a collective identity (see Highmore 2002a: 107–9). As such, Madge and Harrisson's account of the Lambeth Walk is a key component of that configuration identified by Stuart Hall as surrounding 'The Social Eye of *Picture Post*', which prefigured 'the People's War' and 'the high water-mark of the tide of social democracy as a legitimated "structure of feeling"' (Hall 1972: 78, 108). Hall argues the characteristic 'syntax, style and rhetoric' of the *Picture Post* photograph was a 'democratisation of the subject' established right from its October 1938 launch issue, in which a focus on 'participating actors and onlookers' – representative and cross-sectional – 'raises the "unnoticed subjects" to a sort of equality of status, photographically, with the heroic subjects (Prime Ministers) and activities they elsewhere depict' (83). This could equally be applied to M-O's versions of pastoral. Hall argues that the prescience of *Picture Post* is commanding because at that time they 'could not know what such men and women would be called upon to do' (82). Yet M-O preceded *Picture Post* and, as *Britain* demonstrates, were both fully aware of the depth of the crisis at that time and fully conscious of why they were representing the nation and its people in the way they were. There is a strong case for reconsidering the centre of that democratising movement as 'The Social Eye of Mass-Observation'.

This achievement is acknowledged in a displaced manner by Alison Light and Raphael Samuel, in their argument that:

Me and My Girl perhaps captured, or anticipated, something of the mood with which working-class England entered the Second World

War .... It could also be said to prefigure the terms of Labour's 1945 victory; on the one hand, it appeals to a democratic sense of being English, a patriotism which is dependent on a dream of classlessness, whilst on the other, deriving its specific force and passion from a version of distinctly working-class community – an informing belief that it was the working class who were really the heart of the nation (Light and Samuel 1986: 17; reprinted with minor alterations in Samuel 1996: 390–400).

However, when Light and Samuel's article appeared in *New Society* in 1986, it was responding to one of the cultural manifestations of the break up of this postwar settlement: the 1980s revival of *Me and My Girl*. They describe the second half of this production as 'class revenge in which it is the working class rather than the aristocracy who are outwitted' (17–18). Presenting the history of the original production and the Lambeth Walk in general as a refutation of this 1980s representation, they conclude: 'If the point of celebration of popular working-class culture appears so often at the point also of its destruction, then this makes the recognition of the positive potential of those class forms *more*, not less, urgent (18). Yet the point to bear in mind is that the reason why the original production of *Me and My Girl* was so effective politically is because theatre provides an enclosed public space that can be turned into a site of contradiction and contestation – which is to say it can be politicised. To transfer this politicisation to cultural behaviour as a whole would require everyday life itself to be such a public site of contradiction and contestation. This, of course, is exactly the problem with the analysis of the Lambeth Walk in *Britain*: the whole book is predicated on a structural framework that supposes an undifferentiated everyday life.

The consequence of this is not so much to obscure social relations as to enshrine them. In this respect, it is significant that the model of Empsonian pastoral Madge and Harrisson choose to utilise during the chapter on the Lambeth Walk is old pastoral and its implied 'beautiful relation between rich and poor'. Although Empson earlier in the chapter on 'Proletarian Literature' appears to make a distinction between fairy stories and pastoral (Empson 1995: 13) the passage quoted in *Britain* calls this into question: 'The simple man becomes a clumsy fool who yet has better 'sense' than his betters and can say things more fundamentally true; he is "in contact with nature", which the complex man needs to be, so that Bottom is not afraid of fairies; he is in contact with the mysterious forces of our own nature, so that the clown has the wit of the unconscious; he can speak the truth because he has nothing to lose (Madge and Harrisson 1939: 157–8; Empson 1995: 18). Madge and Harrisson continue by adding 'In Shakespeare, the final laugh is usually at the poor man .... In the show [*Me and My Girl*] at the Victoria Palace the situation is rather different – there are plenty of West End people in the audience, but the laugh is really on them (158). This combination of enchantment and performance suggests that the modes of representing the nation developed in *Britain* were – as in the title of Light and Samuel's article – Pantomimes of Class. The threat of Fascism at home was averted not by publicly contesting Everyday Life, but by representing Everyday Life as a site of public theatre.

6
The Mobilisation of Everyday Life

Active Leadership and the Civilian Army

Britain by Mass-Observation came to the attention of a number of the government committees preparing guidelines for the establishment of a MOI at the outbreak of the expected war. The history of the MOI is told in Ian McLaine's *Ministry of Morale*. He suggests that the decision taken in October 1935 to form a subcommittee of the Committee of Imperial Defence to prepare guidelines for the establishment of a MOI on the outbreak of war was a response to Goebbel's Ministry of Propaganda. Its scope, combining coordinated handling of news, censorship, publicity and the collection of intelligence far exceeded the arrangements of the First World War. These roots, probably not so much kept secret as simply unacknowledged, were to prove significant in the subsequent history of the MOI and M-O's involvement with it. On the one hand the system was created to produce propaganda and, on the other, the very secrecy or unacknowledged nature of the undertaking meant that the necessary conscious drive in order to achieve these ends was lacking, while the possibility of it becoming something else – something more than just propaganda – was never really considered within government. Indeed, according to McLaine, the government, especially after Chamberlain's accession, had no real enthusiasm for the MOI, hoping to avoid war in any case.

This became clear with the onset of the Munich crisis, when the MOI was put on standby to go into operation overnight if necessary. Only the straightforward operations such as the news, censorship and administration divisions had a fully planned structure. The two other notional divisions of Home Publicity and Collecting (of intelligence) existed only on paper and as sub-committees. This led to the first

Director General (DG), Stephen Tallents complaining that this was not good enough and criticising the Cabinet for not being sufficiently supportive of the MOI: complaints that contributed to his subsequent dismissal. However, his analysis seems to be borne out by the fact that little more development took place before the outbreak of war while the general indecision and confusion lasted much longer as witnessed by the continuing personnel changes at the top of the MOI – three Ministers and six DGs in less than eighteen months – before it achieved stability in 1941 with Brendan Bracken as the Minister and Cyril Radcliffe as DG.

However, Munich did have a significant effect on the way both Collecting and Home Publicity were to be envisioned. The Collecting Division, in provisional existence since June 1937 under John Beresford, recruited from the University Grants Commission, was getting nowhere. Ivison Macadam, secretary of the Royal Institute of International Affairs – consulted in the wake of the frenzied moments of the Munich crisis – suggested making an approach to M-O. It seems clear that this was in response to the recent publication of *Britain* and associated press articles. M-O had provided exactly the sort of HI that the MOI had as yet been unable to organise for itself. Moreover, the criticism of the press and parliament for not knowing what people were really thinking was especially useful for divisions that were having to justify their proposed existence. It is not surprising that Beresford and John Hilton, Professor of Industrial Relations at Cambridge and Head of Home Publicity, were to become M-O supporters within the hierarchy of the MOI. However, M-O might have had other effects: especially with regard to the way the masses were viewed. A Home Publicity subcommittee had run from July to September 1938 characterised, according to McLaine, by its regarding of 'the British people as intelligent, well-informed and critical' and its recommendation of the clear exposition of regulations and war aims (19–20). The new committee, formed after Munich with the help of the Royal Institute of International Affairs, had a very different set of assumptions to its predecessor, considering that 'information should always flow downwards' and that the masses both 'dislike and distrust argument' (21–2). Of course, this change of approach was directly in response to Munich and had occurred well before *Britain* had been published. However, M-O's analysis of the masses at home, sundered from their leadership by the new media and entertainment industries and prey to the worst kinds of wishful thinking, both reinforced the need for this change and provided some pointers as to how it might be effected.

McLaine reports that discussions took place on 20 March 1939 as to whether M-O should be employed using Secret Service funds – in order to maintain the idea that M-O was working entirely independently and so avoid creating controversy at the prospect of state interference in domestic privacy. At this point, M-O had not been approached but it was decided that Macadam could see them if they passed MI5 vetting (23). From then on, it is clear that there were camps for and against employing M-O within the Collecting Division. In April, Beresford noted the possibility that the Intelligence Division would need to 'produce results somewhat akin to those achieved by M-O' (Beresford 1939). Minutes of a meeting in May, however, record doubts expressed concerning M-O's methods and the likely cost of employing them (MOI 1939a). However, it was obviously agreed that M-O would become part of the Intelligence machinery at the outbreak of war and on 10 June 1939 the Treasury was informed that M-O might be commissioned for various jobs (McLaine 1979: 23, 287).

At the outbreak of war the MOI came into being with Lord Macmillan as the first Minister. Immediately, M-O was asked to report on its first posters and the success of this work led to a payment of £100 for further research (MOI 1939b). Hilton was so impressed with M-O's research on leaflets that he demanded their services be immediately obtained on a full-time basis 'to cover the field of the inarticulate public' (Hilton 1939). He wanted to give them a month's trial for £400 and by the next day, A.P. Waterfield, the Deputy Director General of the MOI, had received Treasury permission for the expenditure (MOI 1939c).

However, the MOI as a whole had attracted strong criticism for the harsh press censorship it imposed and its apparent over expenditure at a time when the prewar financial orthodoxy of balanced budgets was starting to show its inadequacy for the demands of a wartime economy. This meant that Waterfield was told by the DG – Sir Findlator Stewart, who was opposed to the concept of HI (McLaine 1979: 50) – that he had to approach the Minister for his permission on the employment of M-O even though he already had Treasury agreement (MOI 1939d). Thus, on the morning of 26 September, M-O were told that they were going to be employed on a regular basis only to be rung up at 8 o'clock that evening and told that everything was on hold again (Harrisson 1939a). On 30 September, Macmillan refused to sanction the employment of M-O and maintained this position on 11 October in the face of further attempts by Hilton and Waterfield to seek his approval (MOI 1939e–g). By 2 November, Harrisson was reduced to complaining to Hilton that he had still not been paid for all the work on posters and leaflets (Harrisson 1939b). This was to remain the extent of M-O's work for the MOI until the following February. Madge later noted his response to this unexpected financial crisis: 'I took on the writing of a book for Chatto and Windus on our wartime observations, to be finished in a month. This gave me a little ready cash' (Madge 1987: 106).

The very title of *War Begins At Home* can be understood in a number of different ways. The war did start 'at home' through the experience of various emergency measures, such as evacuation and the blackout, long before any British military encounters took place elsewhere. Secondly, it acknowledges that, in the light of the demonstrations of modern aerial power in Abyssinia and Spain, there is such a thing as a 'home front' which is as vital as any other front in the war. However, there is another sense implied by *War Begins At Home* – an imperative carried forward from *Britain* rejecting any notion of retreat into the cocooned domestic world of immediacy:

> All our evidence suggests that it would be easy to transform the present civilian attitude into something even more co-operative, more dynamic and more firmly based on long-term realities – therefore more able to stand tremendous strains which have not yet come, but may come. As President Roosevelt, opening Congress on January 3, remarked, 'It is not good for the ultimate health of ostriches to bury their heads in the sand.' ... We are not ostriches, but some sand-dunes of words and wishful-thought, complacencies, are silting up in a very heady way. We British have not lost our heads in the slightest. We have kept them. But they are also there to be used, and this book tries to trace one possible use which has not been sufficiently exploited – active leadership of the mass of civilians (the majority women)... (vi).

Given that we are promised a follow up volume specifically dealing with leadership in wartime (3), it is reasonable to see this volume as dealing with the led and, specifically, how they can be transformed into the actively led. This context helps to provide the sense of the opening paragraph of chapter one:

> This book is possible because we live in a democracy. And this book will frequently show conflict or confusion between the voluntary principle of peace and compulsory principle of war. This introduction is strictly democratic. It belongs to the voluntary system,

though we hope it isn't out of date! So you don't have to read it;
you can go straight on to the war itself, next chapter (1).

The joke is slightly forced but the message is clear enough: the conflicts
that might appear in this book stem from confusion between former
peacetime democracy and the current demands of war, but the one is
voluntary and the other is compulsory, so there should be no confusion and, hence, no conflict.

In the adoption of this position, Madge and Harrisson broke with
past practice and moved beyond observation to passing judgement,
promising 'if we are alive, to draw plenty of conclusions by this time
next year' (24). This was a definite step towards Malinowski's idea of
ethnography determining a 'correct' theory of society. In accordance
with this functionalist objective, Harrisson and Madge present M-O
as purely a data-gathering organisation: 'We cannot ignore that our
first job is to record and publish sufficient factual data to enable
other students, in other countries and other times, to get from our
work a fair objective picture of what was happening, and to use these
data to fit into their own ideas and re-interpretations' (24). This designation failed to obviate the problem that M-O's collection and presentation techniques were still conditioned to recording the
qualitative experience of the masses – primarily the subjective reports
of their national panel of volunteer observers. The results generated
in such manner could not simply be expected to fit in with the compulsory demands of wartime. This contradiction, not the external
events themselves, is the source of the confusion Madge and
Harrisson identify in the book as coming 'from the fact that the
leaders are badly out of touch with the masses and also out of touch
with the media for influencing, entertaining and explaining to that
80% of the population who left school at 14' (49). The 'bewilderment' of the masses is taken as objective fact and used to demonstrate the necessity of the following assertion:

> A really effectively governed community usually feels clear,
> definite and even violent in certain major issues, quite apart from
> whether its opinions are "true or untrue", "correct or incorrect" ...
> Effective government and leadership has as a prime obligation,
> and also as a necessity for its own survival, the job of telling every
> citizen who isn't in a lunatic asylum roughly what to do and
> roughly what to think about the issues which affect everybody
> (49–50).

Of course, the ultimate purpose of this was to expose the ineffective-ness of Chamberlain and the National Government – a point Orwell recognised in his review of the book (Orwell 2000a: 17–18). However, it flatly contradicts the argument in *Britain* that public opinion was dependent on individuals being given the facts so that they could make up their own minds. This, therefore, was the practical difference between the voluntary principle of peace and the compulsory principle of war.

The prime example given of public bewilderment is in connection with air raids. People were not certain what the different sirens meant and often came out into the street to see what was happening: 'First they have to adjust themselves to the idea of a raid; then they come peeping out to see if it is real; then, as the unreality grows, the nerve-raid fizzles out in anti-climax' (55). This is described as a 'farce' (57), thus begging the question as to what reaction would be considered appropriate: an unconscious Pavlovian response? Surely the people manifested exactly the rational scientific approach M-O had come into being to promote. If experience tells you that most air raid alarms are false, it is only common sense to actually check whether an air raid is real before dropping everything and disappearing into the nearest shelter. Later on in the war, this was to become the standard practice for workers in essential industries.

Madge and Harrisson use their accounts of air raid behaviour as facts supporting a conclusion: against voluntarism and for compulsion. The other factual components of this argument include details of evacua-tion and the carrying of gas masks – down to 2% of the population in Worktown by January 1940 (113). Of course, gas masks were to turn out an unnecessary precaution in the war and, significantly, the argu-ment given in their favour is not the one of major threat to life and limb employed in the case for observing air raid sirens:

> Compulsion and the threat of fine have, in fact, kept the black-out pretty well up to the mark so far. But gas-mask carrying, the only voluntary civilian war-activity in which everyone was recom-mended to join, has dwindled to a minority. As early as October 9, a London park-keeper reports being jeered at by small boys for carry-ing his mask. One important outward symbol of 'unity' has gone by the board (116).

This completely ignores the potential 'unity' symbolised by not carry-ing a gas mask. Orwell wrote a little later in 1940:

As soon as the war started the carrying or not carrying of a gas mask assumed social and political implications. In the first few days people like myself who refused to carry one were stared at and it was generally assumed that the non-carriers were "left". Then the habit wore off, and the assumption was that a person who carried a gas mask was of the ultra-cautious type, the suburban rate-payer type (Orwell 2000d: 197).

M-O's disregard for such oppositional unity is further apparent in their account of the MOI poster campaign:

YOUR COURAGE
YOUR CHEERFULNESS
YOUR RESOLUTION
WILL BRING US VICTORY

This 'attempt to influence mass-morale' is criticised for being 'a failure, by any ordinary commercial standard' because it emphasised divisions (98). A purely quantitative analysis of their research evidence, dividing reactions into positive and negative, might support this conclusion, but their typical qualitative analysis, concentrating on the content of what people actually said, reveals something deeper. For instance, a young man says, 'Your and Us. Never thought of it like that before, but unconsciously that's how I've felt all along: Your and Us' (97). It was not that the posters had generated this sense of division out of thin air. The implied support for suggestions such as the substitution of 'Our' for 'Your' recalls the kind of fantasy resolution of the 'problem' concerning the gulf between the leaders and the led that was so ruthlessly criticised in the 'Munich' section of *Britain*. However, the insight gained from the Lambeth Walk, that everyone could be made to do the same thing at the same time, had convinced Madge and Harrisson that a genuine national unity was possible – nothing less was acceptable: 'It is difficult to run a war in which all the civilian population is so concerned unless every part of the administrative machine is focused in one direction with one idea, with private and public interest fused, military and civil interests fused' (131).

The positioning of the detailed discussion of the blackout after this uncompromising statement is significant. As both the wartime measure that most obviously 'began at home', and the supposedly prime example of how compulsion encouraged unity, the blackout might have been expected to feature earlier in the book. However, in its case, the contradictory currents generated by their qualitative research approach were harder to ignore than in their other investigations. It had to be acknowledged that the blackout was by far the biggest grievance and source of grumbles amongst the imposed war conditions for their national panel respondents and in recorded 'overheard' conversations. Many of the complaints were on specifically political grounds:

> Leeds shopkeeper [overheard]: 'This black-out's quite unnecessary; it's only done to inconvenience the public and make them realise there's a war on' (190).
> 'I dislike the black-out deeply and think it is largely due to a power complex in important people, that it is applied so severely. I find it extremely depressing.' (Staff tutor, 50, Leeds) (200–1).

Harrisson and Madge go on to compare the compulsory blackout for homeowners unfavourably with its relaxation for West-End shops and cars – one screened beam light had to be allowed to cut down on the huge increase in accidents that had ensued from the original restrictions (211–12). Reporting on ideas for improving the blackout collected from the national panel, they note: 'The greatest emphasis in these suggestions is on the need for light in public places' (214). The overall tendency of the account is one of unequivocal opposition: 'On to everyone was lumped this purely negative piece of mass-mathematical compulsion, the black-out' (213). Yet while this evidence contradicted the book's earlier arguments in favour of compulsory measures, it also pointed towards changes in public behaviour that were welcome to Madge and Harrisson. Here, the blackout alone was not responsible, the closure by decree of just about every form of public entertainment at the outbreak of war had completely altered the nature of spare-time pursuits. By November 1939, when the cinemas and theatres were open again, Harrisson and Madge were able to conclude that visiting friends, going for walks and to cinemas, theatres and meetings had all decreased; while 'staying in', reading and going to bed early were the most increased leisure activities. In the theoretical terms which they had long since abandoned, this represented a sharp decline in area 2 activities and an increase of those in area 1. Although Madge and Harrisson had previously criticised home-based attitudes as a retreat from reality, this change was seen as welcome because it provided certain opportunities:

> … the whole structure of British leisure is being changed by the black-out. Much of this change is likely to be permanent. Whereas

there was an increasing tendency for people to go out of their home, drawn by the ever-increasing vested interests of entertainment, this tendency has now been reversed. Home life has gained a new importance. People are being forced to amuse *themselves* more, rather than sit and passively watch others do something. So far no Government and very few private interests seem to realise the huge new home potentialities which can now be exploited (194).

With the new leadership forms of football pools and jazz in abeyance, there was an opportunity to penetrate homelife, which was to come into its own with the onset of the blitz transforming home dwellers into a 'civilian army' – the title of chapter 5 of *War Begins* – performing its duty by going straight into the shelters at the siren. This was the opportunity to erase the common sense scepticism that made people check if there really was an air raid and force them to fully participate in the new national unity.

If *War Begins* had set out, like *Britain*, to map the nation at a moment of crisis, it could have successfully repeated the trick of criticising the Chamberlain government and representing an oppositional national unity in waiting. Instead, it subordinated the inefficiency caused by measures such as the blackout to the supposed symbolic worth of compulsory unity. While the inherent contradictions of both books – caused by trying to fit subjective qualitative material to objective quantitative analysis – are masked by a representational unity, in *Britain* this is linked to popular culture and anti-fascist politics, whereas in *War Begins* it is very much an editorial imposition backed up by the external threat of invasion. As a consequence, 'Mass-Observation' itself becomes a problematic term when discussing the late 1939 to early 1940 period. Does 'Mass-Observation' signify the strident tone of *War Begins*? Or should it refer no longer to Madge and Harrisson, but rather to the volunteer members of the national panel, such as Muriel Green, whose proprietary interest in *War Begins* was noted by us in the introduction to this book. Focusing on the many quoted extracts of observers' experience, rather than the authoritarian editorial soundtrack, allows the book to be read against the grain in the ways suggested above. The oppositional and liberatory sense this generates is reminiscent of M-O's earlier publications. Therefore, it can be argued that *War Begins* marks the point at which M-O, by fulfilling its founders' intentions, actually comes to signify something independent of those intentions: it has become the autonomous voice of a new mass society in waiting. This was the direct, but unintended, result of Harrisson and Madge being forced by Malinowski to adopt positions of individual scientific respectability in line with traditional notions of public authority.

It is the existence of this ambiguity in *War Begins* that makes it so emblematic of the wartime changes *that were to come* – it prefigures how Britain could be both a technocraticly planned society and a site of mass active participation. Symbolically, the book records the transition of the short-lived Lambeth Walk: 'But while these pre-war new dances, which (as we have shown elsewhere) were transforming the atmosphere of the ballroom, simplifying it, socialising it, have tended to decline rather rapidly, new dances are coming up along the same general line of development, with an even further increase of pep and topicality' (229–30). The key example they give is 'Knees Up, Mother Brown' (232). We can see the mode of representation of the nation, developed in *Britain*, in the process of being legitimated by incorporation into an authoritarian framework.

Indeed, even as *War Begins* was being written, Harrisson continued to campaign for the employment of M-O by the MOI. He wrote to his friend Mary Adams – as yet not employed by the MOI – complaining about the 'uselessness' of the MOI in the face of what he considered to be a crisis situation: 'At present, the enormous mass of evidence we have collected suggests imminent collapse of the Home Front, with widespread lack of support largely due to the failure of official and unofficial propaganda (Harrisson 1939a). This correspondence demonstrates that Harrisson was not simply motivated by the need to be employed nor the overriding conviction that the country was about to collapse, but was also involved with a splinter group of like-minded people, such as Adams, Hilton, and Crossman, struggling to change the MOI approach to propaganda and morale. This struggle included a PEP lunch on 1 November with Crossman – then at the MOI – in the chair and representatives of M-O, the BIPO and BBC listener research in attendance. Afterwards, Harrisson sent Adams a memo concerning the 'Co-ordination of Social Research' (Harrisson 1939c).

Meanwhile, there was continued political pressure on the MOI. On 3 October, it had been considered necessary to create a completely distinct Press and Censorship Bureau. As these functions had been the most developed and operational in the MOI, it was not clear what it would do without them and Chamberlain even considered closing it down altogether. In the event, Lord Camrose, Macmillan's assistant, reduced the staff by a third, including the removal of the entire HI division, on grounds of cost (McLaine 1979: 41–2). However, even in this problematic period, events were beginning to turn in M-O's favour. Crossman's tenure at the MOI might have been short – having been signed up at the start of the war to keep the Labour Party happy (McLaine 1979: 33), he resigned on 7 December 1939 feeling that the MOI was not being given a chance to do its job properly (Howard 1990: 79–80) – but while he was there, he completed a report on M-O praising its strength as the in-depth knowledge of specific localities. While he added that caution was needed in accepting Harrisson's conclusions, he recommended M-O's employment for special jobs where its work could be 'invaluable' (see McLaine 1979: 52; Finch 1986: 99; MOI 1939h). M-O's position was further strengthened by Hilton's appointment of Adams to the position of Director of HI – still with absolutely no staff – in December as the MOI began to rebuild itself, a process aided by the replacement of Macmillan by Lord Reith in January.

This renewed possibility of official sanction and remuneration caused Harrisson to send Madge a huge memo asserting the need for 'a really tightly co-ordinated and forceful organisation' (Harrisson 1940a: 1). He argues that there is no proper cohesion between the two sections of the work, which he labels 'Madge' and 'Harrisson', rather disingenuously adding 'no personal allusions are hereafter intended' (3). His view, prefiguring the dominant strains of M-O reception, is that the contradictory currents between 'Madge' and 'Harrisson' extend right back to 1937. Thus, at the beginning, the 'Madge-Jennings approach' had been 'to make a documentary and literary record of mass-life in Britain'. However, 'when mass poetry experiments were unsuccessful', Jennings left 'as the organisation became "too scientific for him"' (4). Thereafter, M-O moved 'towards more firmly-based scientific and statistical work ... [which] in practice has turned out to be the best way of tackling difficult human problems of our civilisation' (6). Harrisson's problem with this development is that Madge remains 'the initiator, the senior member, in the eyes of himself and his friends' (8). Aside from Madge's annoying habit of putting his name first on publications, he continues to preserve the confusion within M-O between 'subjective and objective functions' so that while 'H was building up material for a series of volumes on objective anthropology. M was building up a library of subjective documentation' (9–11).

To this, Madge replied with an equally long memo pointing out that all the Bolton books – none of which had been published – produced under Harrisson's guidance were compiled using indistinguishable documentary methods from those of *May the Twelfth*: 'M doubts if the decisive difference between panel-work and whole-time work can be equated with the difference between subjective and objective, or between the poet and the scientist, or between M & H even' (Madge 1940a: 6). But it is the argument about who was mostly responsible for the published output that brings out the real differences of opinion over *War Begins*. Madge is scathing in his comments:

> Four ... M-O publications have been produced as follows: CM & TH agree to do a book; CM draws up a plan; TH has a counter plan; drafts are produced, mainly by CM (Thus in *First Year's Work*, the section on Pools was pre-boiled and post-boiled by CM, the section on Pub was taken from already written J[ohn] S[ommerfield], the page from Blackpool was a disconnected fragment. In the Penguin, the All-in section was [Gertrud] Wagner. No set of facts was actually written up by TH. Same thing happened with the Chatto Book [*War Begins*].) TH's role in writing up has been
> (a) aggressive criticism
> (b) funnies
> (c) angry passages of own opinions (3).

Harrisson scrawled on the margin of this 'yes but work run by me'. His more formal reply took the form of yet another long memo:

> You are frightened of generalisations? You prefer to produce enormous documentary studies, like the coronation survey. There was a strong tendency, with which you agreed, to turn [*War Begins*] ... into huge blocks of documentation based solely on the panel, (evacuation, black out sections). Such stuff like the pub book is largely unreadable (Harrisson 1940b: 5).

Having already claimed with regard to *The Pub and the People*, 'Sommerfield would be the first to agree that, quite apart from planning the original fieldwork and ideas, I did nearly as much work as he did on the script itself' (3), this can be seen as an acknowledgement that he had been as much responsible for the 'unreadable' literary-documentary style as anyone else. Therefore, by his own admission, Harrisson effectively refutes his claim of only a week earlier that M-O had been riven by contradictions from the start.

If we look at *War Begins* in the light of the evidence from this exchange, the more rhetorical passages can be attributed to Harrisson and the potentially politically resistant chapters to Madge. Nevertheless, this distinction should not be taken as demonstrating that there was an essential disagreement over the aim of the book for while Madge makes clear that the central demand of 'active leadership' for 'bewildered' masses came from Harrisson, he is quick to point out that this is 'a recommendation with which I absolutely agree' (Madge 1940a: 3). However, as he also makes clear, where he does take exception to Harrisson is over this becoming the dominant theme of M-O: '...you base your own drive on a leadership idea which affects the whole organisation of M-O and seems foreign to its first character.' Madge goes on to argue that despite Harrisson's denial of traditional political ambitions, it was readily apparent that the influence he had gained through M-O was tempting him in that direction:

> But there is another possibility (other than M-O being marxist-propagandist or H becoming an M.P.). Namely that M-O should itself become a party, admittedly a peculiar party, but one in which the leader led situations would obtain in a special way, and in which a number of people would be called on to 'believe' in the good faith of one man (it wouldn't work with two, and anyhow it is not M's role). Wartime makes it apparently more necessary to regiment and enforce discipline. A mention on the phone of 'Expelling' from the organisation was symptomatic of a war-time sharpness, while it has been pointed out that the whole-timers are likely to be docile when their chance of being exempted from conscription depends on M-O and therefore on him (8).

Noting that even in Malekula, Harrisson had ended up as a 'policeman', Madge eventually concludes: 'I believe that you believe in the extreme purity of your motives (quite detached in your mind from facading of facts and obsequiousness to public figures, alternating bullying and buttering up of associates) because all that procedure is so highly typical of the politician who could not go on unless he *believed*' (14). Of course, Harrisson's reply hotly denies all this: 'To have one last shot at trying to clarify it. I am political in the same sort of way as H.G. Wells, or Bertrand Russell ...' (Harrisson 1940b: 17).

Considering this exchange, it is possible to suggest that maybe the faultlines in M-O have always been misconstrued – that the differences did not revolve around the science-art axis, but around what kind of movement M-O should be. The tension between the PP strand and a Wellsian 'originative intellectual' movement was already inscribed in the original *Mass-Observation* booklet written by Madge and Harrisson in January 1937. Harrisson had the charismatic qualities of being able to attract support precisely because he could articulate the aspirations of the movement's dominant new-middle-class membership:

> Mrs B ... has asked me about M-O several times ... Tonight she said she'd been reading Friday's *New Statesman* in the train and had read one article which she thought was particularly good and valuable – and so readable and colloquially expressed – had I happened to read it – it was by – now who was it by? – oh yes, a man called Tom Harrisson! Of course, she hadn't connected the name with M-O. She was most impressed when I explained that TH is the head of M-O, and that the article in question was an absolutely typical piece of M-O stuff, and said she now realised what a good idea M-O was and that she must look out for more of it (cited in Sheridan 2000: 96; the article in question is Harrisson 1940m: 300–1).

As had finally become clear to Madge, Harrisson's entire M-O career demonstrated a relentless pursuit of communication outlets from press, radio and television via his courting of the Labour Party – with propaganda based on the Lambeth Walk and the Pools, not to mention stolen Tory canvassing returns – and on to the MOI. At the same time, Harrisson's persistent and heated denials of any interest in 'party politics' suggest that underlying the anguished counter-accusations was a shared anxiety concerning the political consequences of active leadership. This uncertainty, implicit throughout, was openly admitted at least once in *War Begins*:

> Why shouldn't families be conscripted, as well as their heirs? If evacuation is important, why not enforce it? Or alternatively, why not appeal to the heirs to enlist voluntarily, and leave the whole thing on a strictly democratic basis? If evacuation was worth doing, was it fair to leave the decisions to millions of parents who had not sufficient information on which to form sound long-term opinions, nor sufficient leadership-guidance to overcome their personal biases and emotional feelings? (100).

Viewed from today, one would have thought that the obvious point of comparison for Madge and Harrisson in studying the emergency measures introduced at the outbreak of war, would have been their own account in *Britain* of the ARP mobilisation during the Munich crisis. They reported this at the time as bringing fear into the home, so exacerbating the crisis and allowing Chamberlain's third flight to Germany to become a fantasy resolution. A simple connection would have seen the September 1939 measures as equally inducing fear in the home, exacerbating the crisis and distracting potential criticism of the government's general inaction and failure to attack Germany on the Western Front. Madge and Harrisson could then have been expected to criticise the official measures as fantasy and show the everyday scepticism of the public as a refusal to be duped. In which case, they might just as easily have marshalled their contradictory evidence in favour of voluntarism as the best means of achieving national unity – an argument that would have been subsequently born out by the quarter of a million men who responded within the first twenty-four hours to Eden's appeal for Local Defence Volunteers (Calder 1992a: 105).

Paradoxically it was exactly their experience of the Munich crisis that caused both Madge and Harrisson to come down in favour of compulsion. They knew from that study that public opinion had not been behind Chamberlain, but they also thought this detectable only from their own qualitative techniques. With no belief in a public sphere, they could not conceive of anti-Chamberlain public opinion being expressed during the phoney war unless it was through 'active leadership'. It was the utopian dream of *Britain*: leaders speaking for the led with one voice, speaking *for* Britain. As such it tuned in with the wider consciousness of the time, most publicly expressed during the House of Commons on the evening of 2 September 1939. Faced with Chamberlain's apparent continued belief in the possibility of peace, and the prospect of a 'second Munich', Arthur Greenwood, the acting Labour leader, rose to speak and was greeted with the shout of 'Speak for England, Arthur' from the Tory, Leo Amery. Other Conservatives took up the call while Labour members shouted 'Speak for the workers' and 'Speak for Britain' (Calder 1992a: 33).

The Ministry of Everyday

From the beginning of Adams's appointment as head of HI, she was keen to employ M-O. No doubt this was in part because she had no other means for gathering intelligence at her disposal as many of the government network of RIOs had been sacked in October. She pointed out in an early 'Memorandum on the Function of Home Intelligence' the need for independent investigators, suggesting that the first requirement was to decide the role outside agencies should take, only after which would it be clear what type of personnel would be required (Adams 1940a). In a subsequent memo, she proposed creating a 'morale barometer' and a follow-up to the Southwark by-election amongst other ideas. Page two of the memo suggests that M-O were seen as the potential fieldworkers for the barometer in association with headquarters staff's analysis of other available sources of opinion such as postal censorship, MP's postbags and the RIO network (Adams 1940b).

In a memo to Macadam dated 20 February, she recommended reading *Us* – the M-O weekly newsletter which had just been started by Madge against Harrisson's advice – and *War Begins at Home* (Adams 1940c). A letter the same day from a Mr Crutchley of the Ministry of Home Security – part of the Home Office – suggested that, despite qualifications, M-O's *War Begins* was useful and interesting (MAP: 2/E). All these elements must have combined into a configuration favourable for the employment of M-O. The next day, Adams phoned Harrisson, and then wrote formally, to commission a report on the upcoming Silvertown byelection to be conducted under the name of M-O with no mention of the MOI. The special brief was to investigate apathy and feelings about the war, which caused 'anti-war feeling to find expression in support for the Fascist candidate on the one hand and the Communist candidate on the other' (Adams 1940d). On 5 March, Adams commented on M-O's Silvertown byelection report: 'Since this was the first investigation undertaken for me by M-O I made a point of checking up on certain observations and am satisfied with the objectivity of the results' (Adams 1940e). Three days later, she organised for M-O to be employed for £60 to report on the Leeds by-election (Adams 1940f). On the same date, she also submitted a memo giving the history and her analysis of M-O to Macadam (Adams 1940g). On 12 March, she wrote to Waterfield arguing that M-O should be employed permanently on the same basis as had been suggested to the Treasury during the previous September (Adams 1940h). From 26 March onwards, Harrisson began negotiating – as to what rights M-O would retain – an agreement, taking care to point out in somewhat pained tones that due to their experiences of the previous autumn there was a 'certain amount of reserve on our side to commencing any research until it has been finally and fully sanctioned from your end (Harrisson 1940d). On 1 April, Adams requested monthly expenditure of £1,500 and £400 respectively to set up statistical surveys and employ M-O (Adams 1940i). This latter request was clearly formal, presumably following permission to employ M-O from the DG in response to the earlier memos. The setting up of what would become the WSS was partly to act as a check on M-O and thus in keeping with the idea of maintaining an air of caution with respect to M-O.

On 8 April, Harrisson stated his understanding of the terms of the contract with the MOI: including M-O's continued independence, the right to do its own work aside from the MOI's, publication rights of all work including that done for the MOI, continuation of *Us*, petrol allowance, military exemptions and a refusal to undertake 'espionage' (Harrisson 1940e). The MOI agreed, with the provisos that publication be conditional on their permission and that they would try their best with exemptions (MOI 1940a). On 16 April, the expenditure having been agreed for an initial three month period, Adams requested that M-O's payments should be sent weekly (i.e. £100 pw), commenting 'M-O began working at the agreed pressure from April 6' (Adams 1940j).

With both funding and independence assured, Harrisson could feel secure enough on 10 May 1940 – the day of Chamberlain's resignation – to use the pages of *Us* to take stock of public morale following the failed military intervention in Norway. Linking the complacency and wishful thinking of the phoney war with the abdication, the coronation and the Munich crisis he suggests that a sense of unreality fed by false press reports had grown and grown until the defeat in Norway had led to an explosion of doubts: 'It is not a political upset of people, but an upset about information, facts, authenticity, the external world of foreign events which has always been vague to the mass of the population who left school at 14. It isn't anything to do with defeatism or being disheartened' (Harrisson 1940g: 151). From 18 May, as the German armies moved across Holland and Belgium, M-O began providing the MOI with a sequence of daily morale reports gathered from around the country which continued unbroken to 12 July. These were supplemented by Harrisson's various ideas about propaganda: 'Home propaganda should be made interesting like *Picture Post*' (Harrisson 1940f). In particular, he advocated a 'middle path' between the 'two extremes' of the 'democratic way' and the 'totalitarian way' (Harrisson 1940h: 23–4). This idea culminated in a 'Yardstick Memo' providing a graded scale for effective propaganda in which firm statements such as 'we are fighting for civilisation' and 'we are right' are contrasted with the negative effect of 'we don't have war aims' or even the publication of unemployment figures (Harrisson 1940i). By this time in June, Harrisson was sitting with the Morale Advisory Committee, which included friends like Crossman, Julian Huxley and Tom Hopkinson of *Picture Post* (see Adams 1940k). The daily morale reports of this period are characterised by repetitions of Harrisson's fundamental concern that the gulf between the leaders and the led was continuing to widen. It is noticeable that Harrisson did not see the Dunkirk evacuation as a turning point and that he treated calls for revolution as examples of further public 'bewilderment':

> This situation, so closely similar to the position in early May, bears all the marks of a potential explosion in the public mind. There is a very serious danger that this explosion might be too extreme for rapid readjustment. The whole prestige of leaders has so seriously declined in the past few weeks ...
>
> On the one hand the danger is that the latent and faltering move-ment to drag down Chamberlain and his associates might drag down everything. On the other hand, the danger that the continua-tion of Chamberlain [in the cabinet] against the mass democratic wish, will produce mistrust and suspicion and so rot popular inter-est so much that this too will drag down everything (Harrisson 1940j: 5).

In early July, Adams wrote to a Mr Welch of the Treasury that 'I have no hesitation in saying that Mr Harrisson's reports have been of the greatest value, and have now become an essential part of the whole machinery of Home Intelligence.' She went on to request that M-O's contract be renewed for a further three months (Adams 1940l). Welch's reply on 6 July was the first sign that financial problems were going to cause problems for Adams and M-O in their work for HI, as he stated that it was not possible to authorise such a continuation of the contract, although payment could continue for the present pending further consideration and reports (MAP: 4/E). Harrisson's response to this unwelcome development was to revive *Us* – which had been allowed to lapse – in a mimeographed format as a guaranteed vehicle for his views on morale: 'every internal attack on our leadership, however "necessary" from the political point of view, may have an unsatisfactory mass effect' (Harrisson 1940k: 6). He goes on to com-plain: 'It is so difficult for those of us who are intelligent to remember that for a large part of the population a lot of the subjects which agitate our moods are utterly obscure or non-existent.' The sense of frustration is palpable as the escalating financial crisis at the Treasury threatened to undo M-O's hard won position, with HI coming under pressure from the DG to justify its expenditure once again.

On 17 August, Adams submitted a report on HI's work under the headings 1) WTSS, 2) M-O, 3) HI Daily Reports (Adams 1940n). The M-O list includes Madge's work on saving and spending in Coventry – which, as discussed in the next section, was not being funded through the MOI – suggesting a desperate attempt to make the operation look as big as possible. By September, Adams was forced to formally notify Harrisson that financial authorisation for M-O work was due to expire on 13 September and that while she hoped to extend this for a month, he must be prepared for a termination of contract on 13 October subject to the findings of a review by the new DG, Frank Pick (Adams 1940o). Pick, by this time, had already put the Morale Committee in abeyance. Harrisson wrote back on 5 September, expressing upset: 'It had seemed to me that there was a growing appreciation of the value of this work, I gained this impression from many people within the Ministry and in other Ministries' (Harrisson 1940l).

The severity of the situation is made clear by the memo from Adams to Macadam concerning the reorganisation of the WSS: 'As you know everybody in the employ of the Survey was given a month's provisional notice on September 10' (Adams 1940p). On 26 September, a further memo noted that 'financial authorisation for WTSS and M-O expires on October 13th ... It is essential therefore, if the DG approves of the scheme of re-organisation we have put forward, that Treasury sanction for future expenditure be obtained at once'(Adams 1940q). These proposals entailed a 40% reduction in expenditure although M-O's funding was to remain at the same rate as before. On the next day, a meeting took place, with Pick in the chair, that substantially restructured HI's work:

Daily Report on Morale by Home Intelligence

In view of the decision of the War Cabinet, this Daily Report will cease forthwith. In place of it two reports will be produced by Home Intelligence as follows: –
A daily statement of facts requiring treatment or likely to affect public opinion which have been brought to the notice of the Ministry.
A weekly report on changes in public opinion and habit as reported to the Ministry by selected and specified sources.
....
to be without comment.
This document will be used as a guide for action by the departments of the Ministry concerned with publicity at home. It shall not be circulated or made available to anyone outside the Ministry of Information (MOI 1940b).

While M-O and the WSS still remained at joint number one on the list of allowable sources the circulation of their findings was being severely curtailed and their scope for comment and interpretation was also being narrowed. In any case, the subsequent weekly reports show through their list of sources that M-O was rarely used in practice for this particular purpose, but rather for special reports only – the circulation of which was equally curtailed. It seems that the opportunity provided by the necessity of restructuring due to the national financial crisis was taken to impose some political controls. By that time in September, the Cabinet would have been already aware that Hitler had called off the invasion of Britain and that it had a period extending at least until the following Spring in which to deal with the domestic situation. The restriction of HI reports to purely propaganda purposes was a way of restricting information about public opinion which was in favour of a more radical pursuit of the war than the executive was prepared to allow. This was a watershed in the history of the MOI following the five months in which things had developed according to the way Adams and Harrisson had desired. A number of brief memos from Adams to friends and contacts state her feelings in various phrases such as 'we're having hell here', 'muddle', 'strange developments here' and 'what a stupid life this is' (see MAP: 2/E).

Harrisson's response to this increasing marginalisation was characteristically robust. Clearly believing that the maintenance of morale was more important than ever at a time when London was being bombed nightly, he made sure that his views received the maximum possible circulation by having them published in the 28 September issue – the same issue mentioned in the M-O diary extract quoted in the previous section of this chapter – of the *New Statesman*: 'Our problem is to adapt ourselves to the new circumstances of bombardment. Here I am discussing three aspects of this adjustment – domestic arrangements, leisure and sleep' (Harrisson 1940m: 300). Discussing a range of issues from the standards of public air-raid shelters to the huge increase in time 'being spent negatively at home', the article concludes:

> This confusion is not *only* due to the confusion and lack of co-ordination between authorities. To deal with that, with the immediate problems and past mistakes, we need a Welfare Board ... But we need also to go further and to *look ahead*. Welfare is something additional, sedative, supplementary. We need now a group, a special branch of Government, to deal with the whole human problem .... Many London problems come from the attitude of people to authority, to official instructions, to fantasies about danger as well as sens-

ible fears. To cope with all this, we need a wideawake unit, composed of ordinary sensible people (not knights or professors). Their
job: to foresee the difficulties caused by the human factor .... The
time has come when we must see that mental upsets and social
conflicts can be more serious than destroyed property or physical
casualties. It should not be possible for London to be robbed of
sleep for a month, and still no 'sleep policy' put forward by
Government. Perhaps we need a whole new ministry. The Ministry
of Everyday (301).

It was presumably not coincidental that two days later Adams
received a curt memo: 'What does M-O do for £100 [per week]? How
many reports have we received during the time we have been paying
them at this rate? How is the figure computed?' (MOI 1940c). She
obviously managed to deal with the enquiry and preserve M-O's vital
stipend but as M-O was no longer used as a major source for regular
morale reports, it was necessary to find special projects for them. On
14 October, Adams quizzed Harrisson on whether pay-day should be
changed to Thursday to help housewives get the shopping in before
the weekend (Adams 1940r). In November she wrote to Harrisson
asking him to concentrate on attitudes to democracy, to Allies and to
Russia (Adams 1940s). In the meantime, it proved impossible to get
Harrisson recognised by the Stamp Committee as a Social Science
Research Worker and so he had to rely on renewing temporary
exemptions from conscription. The irony surrounding who was and
was not awarded Social Research Worker status had not escaped
Adams – or Harrisson – as she had made clear back in August: 'It is
interesting to note that Mr Charles Madge has been given exemption
because of his identification as a social worker by the Stamp
Committee...' (Adams 1940m).

It soon became clear that Adams was waging a rear guard action. Her
hopes were pinned on Sir Walter Monckton, who took over as DG in
the new year. In a memo to him she stated her case resolutely:

Behind all present and past difficulties is the question who is
responsible for civilian morale?

Is it MoI, or Home Security [a branch of the Home Office] or
Cabinet.

Work carried out at the moment is limited in scope ...
[partly because]
... The Minister has never considered it proper to make use of our
reports in any policy decisions that affect morale (Adams 1941a).

She also noted with respect to M-O: 'Personnel situation becoming increasingly difficult. All personnel of military age gradually being called up. Difficulty of getting exemption.'

Adams wanted HI either to get out of the MOI and provide a service for any interested Government departments, or to be properly integrated in an MOI structure that sent reports to the Cabinet. Instead, all the material they collected was being used for press direction. At the same time, HI was still recruiting staff: correspondence in February related to the employment of Richard Titmuss and Cecil Day Lewis. Despite this recruitment activity, Adams was still unhappy, writing to Harold [Nicolson?]: 'This wretched place is undergoing yet another reorganisation.' She went on to complain that 'the trouble is right at the top' and reiterated her desire that HI should be free of the 'vulnerable' MOI 'in some other part of the Administration where access can be direct and without prejudice.' She also recorded the 'terrific pressure everywhere to <u>say</u> (especially in cold print) that everything is alright, that morale is splendid.' Her belief in the validity of HI fieldwork specifically included M-O (Adams 1941c). She also wrote to Julian Huxley that 'never in my wildest dreams had I believed such inefficiency and inability exists'. This was clearly directed at Monckton, whose policy she encapsulated as 'delay, delay, delay' (Adams 1941d). On the same day, a letter from Harrisson stated. 'I am with you all the way in any tough line that you take.' He went on at length:

In my humble opinion, the files of your department contain vast quantities of material, which if suitably circulated, supported and sponsored, have very greatly assisted in reducing the appalling chaos and unnecessary misery which is piling up on the civilian front, and which is rapidly accelerating in a downward direction.

I am behind you in anything you do and am not in the least concerned about our own security etc. All of us concerned with Intelligence work have too long held our own security in pawn to Home Security.

... any arrangements of this sort [i.e. M-O's work for Naval Intelligence, which was just beginning] which I may make are subject to understood revision whenever you are in position to discharge an efficient service to all Ministeries without interference...

... Let not our new battle cry be: "Parker [the recently appointed new Director of Home Publicity] must go." Go somewhere with some *Liebestraum* and leave after the end of March, if you cannot take it before! And on the way out, let us take Ling-fish and drop him in the slime (Harrisson 1941a).

While she was able to write on 7 April 1941 to Harrisson that M-O's contract had been extended for another six months on the existing financial terms, terminable at one month's notice on either side (Adams 1941e), by 9 April an office circular was announcing her retirement (MOI 1941a). She did not entirely abandon M-O, trying to ensure, as she worked out her notice, that someone was assigned to write directives for them and to ensure no overlapping (Adams 1941f). At this point, HI ceased to be a separate division but became a sub-section of the Home Publicity division, under Stephen Taylor who had previously worked for Adams but who was a sceptic about the value of M-O. M-O's contract was finally terminated on 30 September 1941 (MOI 1941b). Afterwards, Taylor wrote to Harrisson thanking him for his past work and reminding him that he needed written permission to use material that M-O had gathered for HI (MOI 1941c). Harrisson replied by saying 'I believe that I could have been a lot more useful given more liaison and directives' and he went on to ask for *ad hoc* jobs (Harrisson 1941b). Later in the month, he offered several topics which Taylor rejected (Harrisson 1941c). In November, M-O was employed to report on the ATS campaign which appears to have been their last main collaboration with the MOI (MOI 1941d). Thus M-O worked for the MOI for the first two years of the war, but only under a full-time contract from 6 April 1940 to 30 September 1941.

Ironically, it is possible to show in retrospect that Harrisson's *New Statesman* article represented a greater success in bridging the gap between the leaders and the led than any of M-O's work for the MOI. When the government files for 1940 were made publicly available on 1 January 1971, it emerged that Churchill, himself, had read Harrisson's article and brought the matter up in the war cabinet meeting on 2 October 1940. Referring to the description of conditions in a shelter in Stepney by an 'East End working girl', he called for 'drastic action' and the removal of the Stepney ARP Officer (Crossman 1971: 176). The conditions in question demonstrate a fundamental disunity at the heart of what was to become the Myth of the Blitz:

> The structure is colossal, mainly of platforms and arches. The brick is so old, the place so filthy and decrepit, that it would be difficult for the best of artists to convey its ugliness on paper. First impression is of a dense block of people nothing else. By 7.30 p.m. each evening, every available bit of floor space is taken up. Deck-chairs, blankets, stools, seats, pillows – people lying on everything, everywhere.

When you get over the shock of seeing so many sprawling people, you are overcome with the smell of humanity and dirt. Dirt abounds everywhere. The floors are never swept and are filthy. People are sleeping on piles of rubbish. The passages are loaded with dirt. There is no escaping it. The arches are dank and grim – they are lighted well, until black-out time, and then all the lights on one side are put out, because the black-out arrangements are inadequate.

There they sit in darkness, head of one against feet of the next. There is no room to move, hardly any room to stretch ...

The sirens went at about 8 p.m. Lots of people were asleep, but in general it was too early for this.

But already, at 8, people were beginning to cough, and this coughing spread, and lasted throughout the evening. I developed a cough and sore throat, in the early stages of the evening.

Everyone there was working class. The shelter is near the dock area, and near the coloured quarters. Mostly Cockneys, but also many Jews and Indians. On the whole, the Jews lay on the right-hand side, the Cockneys in the middle, and the Indians on the left. Race feeling was very marked – not so much between Cockneys and Jews, as between White and Black. In fact, the presence of considerable coloured elements was responsible for drawing Cockney and Jew together, in unity against the Indian. There were a lot of cases of mixed marriages – in fact, it was more usual to see a mixed one, than to see husband and wife coloured. Some of the coloured people were Indian, some Negro, a few Chinese. Some of the Indians, those not occupied with girls, played cards (cited in Harrisson 1940m: 300–1).

Harrisson prefaced this extract with a statement that was simple but resonant: 'It is worth while, at this moment of history, putting on record a simple description of one of the few deep shelters in the East End.' The preservation of such eyewitness accounts is ultimately the greatest justification for Harrisson's approach. However, at another level, the whole episode serves to demonstrate the fundamental limitations of his strategy at the time. By bridging the gulf between the leaders and the led, he merely triggered a reflex response. The net result was to make the 'additional, sedative, supplementary' approach, he associated with 'Welfare' more effective and thus undermine his wider objectives of changing the whole way that society functioned.

Co-operating with the Tax Collector

It is possible to project an entirely different trajectory for M-O's wartime work over the crucial 1940–1941 period. On 17 April 1940, Madge wrote to John Maynard Keynes: 'The Ministry of Information work only involves the unit working under the direction of Tom Harrisson. I am not involved in any way, nor is Mass-Observation as an organisation ... I'm personally very glad that M.O. will be working for the National Institute [of Economic and Social Research (NIESR)]' (cited in Keynes 1983: 814–15). As the economist Donald Moggridge has noted, by playing a role in the formulation of the 1941 Budget, this collaboration represented 'the earliest use of social survey techniques in the formulation of fiscal policy in Britain' (Moggridge 1992: 644n). In order to expand on this footnote to history, it is necessary to first look back at the Worktown economics project that ran from November 1938 to August 1939.

An unsigned M-O memo under the heading of 'Social Factors in Economics', which refers to the importance of investigating the confusion between economic and social-personal needs (M-O 1938b), marks a significant step in an ongoing attempt to drag the British social survey tradition out of the nineteenth century. Following the influence of the Marienthal survey, research projects into unemployment such as Oeser's investigation of the Dundee jute workers and the work funded by the Pilgrim Trust which resulted in *Men Without Work*, had incorporated the perspectives of social psychology. Now M-O wanted to first revise and then apply these techniques to investigate the social psychology of saving and spending across the working class in general. To this end they assembled a strong unit including two scientifically-trained personnel with relevant experience, who already had connections with M-O: Chapman and Wagner. Chapman had worked with Oeser at Dundee and, earlier, as a research assistant on Rowntree's repeat survey of York. Wagner, who had been left in control of 85 Davenport Street following the completion of the first Worktown phase – in which she had carried out the research into all-in wrestling that appeared in *Britain* – at the end of August 1938 (Barefoot 1939: 201), was a veteran of both the Marienthal and the *Men Without Work* surveys. Chapman would go on eventually to join the WSS, where he recruited Wagner and two other M-O workers on the economics project, Geoffrey Thomas and Kathleen Box. Thomas would go on to become head of the GSS, the peacetime successor of the WSS, in the 1970s. Moreover, as Liz Stanley has demonstrated authoritatively, the economics project was at the centre of a number of networks linking M-O with established academics:

> ... there seems to have been many connections between various of the Worktown researchers and the LSE [Arthur Bowley, Arthur Carr-Saunders] ; fewer but still important connections between the Worktown researchers and the Department of Economics at Manchester University [John Jewkes, Hans Singer] and a subsidiary one with the psychology department there under T.H. Pear; at this time indirect connections with Phillip Sargant Florence and Wilhelm Baldmus at the University of Birmingham; and a possible research student connection with Percy Ford at Southampton University. Later after the war, there were strong connections between former Worktown researchers and the Universities of Liverpool and Birmingham' (L. Stanley 1990: 26–7).

Therefore, the project was under a certain amount of pressure from the start – not least because Gollancz had yet again provided funds in the form of an advance for what was expected to become the 'fifth' Worktown book (Madge 1987: 84). This pressure affected the M-O founders in different ways. On 15 November 1938, Madge wrote in a letter: 'I have a houseful of bloody experts here now, and I foresee that M-O is going to be a series of skirmishes and battles between them and me and Tom' (Madge 1938d). Harrisson, however, was worried about Madge and Wagner combining against him, as he wrote to Chapman: 'I'm looking for you to counteract the somewhat academic tendencies of Wagner and Madge' (Harrisson 1938d). Chapman, who arrived in Bolton after the others and was only to stay for six weeks, was angry because Harrisson had promised to pay him £35 for his time. This in turn angered Madge: 'One [of the people here] was promised a lot of money by Tom and it is a month overdue and he is very disgruntled and is making trouble with the others: either I shall have to squash him, or else kick him out. But the difficulty is that when I go away, things tend to be slack ...' (Madge 1939a). This private letter shows the hard side of Madge that was rarely on public display. Despite the common perception that Harrisson was the dominant personality of M-O and Madge the 'loyal opposition' within (see e.g. L. Stanley 1990: 16, 32), there is an overwhelming case for arguing that Madge was always in control of whichever part of the organisation he was currently involved with. As we have seen, his name went first on the three M-O books he completed with Harrisson and he continually instigated projects in the name of M-O from *May the Twelfth* to *Us* against Harrisson's judgement. Crucially, Madge's own understanding was that he was in charge:

> I came to Birmingham yesterday to meet Tom after a remarkable exchange of letters, the offshoot of which has been to re-establish my dominance over him, which was disappearing. But the funny thing is that in his letter he was pleading to be dominated again, which is a very revealing indication of his psychology.
>
> ....
>
> I have always found that if I exerted my own personal influence on ANYONE, they caved in. It is a rather dangerous power and one which I keep switched off most of the time, so much that I give most people the effect of mildness (Madge 1938f).

However, even though the male pecking order was eventually established, nobody thought to tell Wagner of this. As Thomas noted in his diary for 11 February 1939: 'Gertrud says, what I want is this, what I suggest is this, what I want to get is that – poor Charles is only secondary in her estimation' (G. Thomas 1939). Indeed it was Wagner, who went on to write up the fullest account of the project as an MA thesis, 'The Psychological Aspect of Saving and Spending', submitted to the University of Liverpool in October 1939. In her thesis introduction she criticises *The Study of Society* and, in particular, Oeser's theory of functional penetration:

> It seems to us that the method of 'functional penetration' was adapted because the investigator felt he had to use hidden means in a time when the right and necessity to carry out investigations in the field of social psychology are not recognised by the agencies of society. The question arises whether we are not near the point where the social psychologist should come forward into the open with his demand for investigation of social behaviour and whether public opinion could not be convinced by an appropriate campaign that help in such investigations would benefit society in the end (4–5).

The 'us' here clearly refers to M-O and, indeed, her study is based on two groups: the Bolton working class and 'mass observers (more appropriate name would be self observers) from the national panel of Mass Observation' (1). Wagner goes on to explain that she conducted her research by lecturing on Austria – a topical subject at the time for obvious reasons – to different organisations and institutions, in order 'to win the personal sympathy and interest of the audience' so that when at the end she explained her current investigation, stressing the benefit to the community, she 'got invariably more addresses than she could actually make use of' (5). The clever part of her method was based on the fact that the interview generally took the form of a big Lancashire tea: 'During tea the investigator was usually able to collect all necessary objective information about the income, the rent paid, hobbies etc. so when the actual interview began, she was already clear about the background of the person interviewed' (6). This helped in counteracting 'the very unfortunate desire on the part of the informant to give a logically fully proved answer' because the interviewer, by dint of what she had already learned could guide the conversation to ensure that 'the informant is encouraged to tell everything he can think of in connection with the subject.' The introduction concludes with a reformulation of a classic M-O principle: 'Every action takes place in a form which is more or less strictly prescribed by the unwritten rules of the community, and it is the task of the investigator to find those rules out' (8).

Chapman later insisted that Wagner's approach was more indebted to Oeser than she allowed:

> Wagner was academic in the sense that she had an academic training and she was a psychoanalytically trained psychologist and was interested in the emotional substructure of working class life and depended very much on Bill Naughton as her key informant. Bill Naughton from delivering the coal in Davenport Street became the person on whom she chiefly depended to get information about the families that he visited. He was then, I suppose, strictly a participant observer, but what Oeser would have called a functional penetrator of a social field because his function as a coal delivery man gave him an entrée to a social field that would be closed to any other person' (Chapman 1979a: 6).

However, Chapman's difficulty in defining Naughton's role encapsulates a wider problem of definition as the identities of 'ethnographer' and 'native' blur. What this really illustrates is how much this research remained true to the moment of mutual recognition at the heart of the first phase of the Worktown project. Therefore, it is impossible to fully agree with Liz Stanley's conclusion that the economics project represented a commitment to 'radical quantification' (34). Undeniably, many of its participants went on to work in the positivist, survey-oriented postwar world with considerable enthusiasm for planned solutions to the problems of society, but these careers were based on taking a public professional stance, which obscured the personal interactions which invariably underlaid them. Wagner's MA thesis is a case in point because it obscures her considerable personal links with Worktown even while repeating such platitudes as: 'One of the most important qualities of a fieldworker is to lose all emotional attachments to the problems and groups he is investigating' (Wagner 1939: 4).

Likewise, Madge's later comment in an interview with Calder that he eventually became frustrated with M-O because 'one couldn't get good solid quantitative results from it' (Madge 1979: 18), implies that his move to this kind of work with Keynes and the NIESR was a big break from M-O and its messy qualitative techniques. Yet it started with this economics project, which stubbornly remained embedded within M-O's complex relationships with the local community. Indeed, when Madge first contacted Keynes in April 1939, he boasted about exactly this aspect of the research:

> Our approach is as personal as possible, since working-class people react strongly against any 'official' inquiry. Things have not altered much since 1833, when the questionnaire sent out by the Factory Commissioners on working-class savings was such a failure. But our team has been living here for over two years, our contacts are wide and varied and the material we need comes rolling in (Madge 1939b; also in Keynes 1982: 519).

Keynes was interested in 'Social Factors in Economics' and expressed interest in the results but in the end, although the research was mostly completed, it was never written up as the intended book because the war broke out and priorities changed.

However, after falling out with Harrisson, those priorities no longer seemed so important and Madge was struggling with the costly and unsuccessful business of bringing out *Us* every week. It was at this time that he read Keynes's *How to Pay for the War*, and was inspired to write to him again proposing a research plan that would help solve this complex problem. Keynes was convinced of the need to transfer resources from peacetime to wartime uses in general, which meant reducing peacetime levels of consumer spending in particular. His proposals were that this should be achieved by increasing tax and enforcing compulsory saving, which would be credited as deferred pay after the war. However, the Treasury remained stubbornly wedded to the *laissez faire* controls of voluntary saving and inflation, while the trade unions supported neither of Keynes's proposals. Therefore, not only was Keynes interested in Madge's ideas for finding out exactly how the war had affected people's saving habits, but also in his offer of discussing how the plans could successfully be 'put over' to the 'mass of people' (Madge 1940b; also in Keynes 1983: 810–1). While it is not known exactly what was discussed in the meeting between the two that took place on 21 March 1940, it is fairly easy in retrospect to see exactly what Madge's value to Keynes was. In order to prove that the graduated tax system he favoured would work better than a flat-rate tax, Keynes had to demonstrate that the working class would accept being brought into the tax system for the first time (see Moggridge 1992: 630–1, 642–7; Calder 1992a: 237–8). Therefore, Keynes needed information and Madge was one of the few people with the necessary experience and personnel to be able to devise, carry out and report on fieldwork surveys of working-class areas. The outcome of the meeting was that Keynes gave fifty pounds of his own money to Madge to get him started and recommended M-O for subsidy to the head of the NIESR:

> You probably know by name at least the two lads who have been making themselves responsible for what they call Mass Observation – Tom Harrisson and Charles Madge. In my opinion they are live wires, amongst the most original investigators of the younger generation and well worth encouraging (Keynes 1983: 812).

On the strength of Keynes's word and Madge's outline memo for research, the NIESR council circumvented the minor problem of ineligibility arising from M-O not being part of a university by attaching them directly to its own headquarters and the new collaboration began.

By coincidence, M-O had been taken on and funded by two different official bodies in a matter of days. The consequent reorganisation was announced in *Us* on 26 April: 'Two whole-time units will work under separate direction' (M-O 1940: 121). Harrisson was described as concentrating on morale and opinion formation in London and Worktown, while Madge directed a mobile unit investigating social economic factors in a programme of intensive studies at a series of different locations. The first of these locations was Islington and the second was Coventry. From where Madge wrote to his mother:

> Coventry is full of wonderful brand new pubs, cinemas, stores and banks, with a nucleus of timbered houses and spires. It is really striking to walk through street after street of working-class homes with gardens a mass of roses, honeysuckle and philadelphus perfuming the air. In many of these streets every house has a garage. The great majority are smartly dressed (cited in Madge 1987: 123)

It is impossible to read this today without reflecting on the subsequent devastation of Coventry in the air raid of 14 November 1940. Up until that point, Coventry had been – as Madge's letter testifies – a 'boom town' (see Calder 1992a: 29). It typified the emergent classless England recognised by Orwell in *The Lion and the Unicorn*. Indeed, even after the bombing, Coventry remained a major centre of war production and its continued affluence was to attract official investigation later in the war (see Calder 1992a: 219, 453). It was exactly this concern about earning levels being too high that had brought Madge to Coventry in 1940 as the letter to his mother confirms:

> Trade is brisk. National Savings are doing well. But when all the spending and all the saving is taken into account, there still remains a gap between this and total earnings. My suspicion is that much cash is being hoarded in pockets and drawers – the very worst thing from the point of view of the economist. I am happy our enquiries will settle the point (cited in Madge 1987: 123).

There was a certain irony at work here: in fact, the situation in Coventry was the very best thing because it confirmed the point of view of Keynes. Madge's grant with the NIESR was renewed, but this time he decided to hold it as an individual. This seems to have been because the combination of invasion threat and financial uncertainty at the end of June had resigned him to having to join the army, and this sudden reversal of fortune suggested to him the opportunity of a new beginning (see Madge 1940c). He went to Blackburn next, from where he noted the discomfort caused to the MOI by press attacks:

> Perhaps they will get the sack. I am thankful not to be involved.... There is a stupid letter in *Picture Post* signed 'Tom Harrisson, Mass-Observation' saying that it is the duty of leaders to lead, that it

doesn't matter what the masses think and that Churchill is right in not yielding to popular demand that Chamberlain should go – a very silly sort of letter ... He appears to be rapidly going down the drain (Madge 1940d).

Keynes greeted Madge 'like a son' after seeing the figures on his return from Blackburn (see Madge 1940e) because at last he had evidence to persuade the Treasury to his way of thinking. Ample funds were now made available to confirm Madge's employment for the foreseeable future, so that the good work could go on and Madge's report on Coventry, 'War-time Spending and Saving', and the subsequent report on Blackburn and Bristol, 'The Propensity to Save in Blackburn and Bristol', were published by Keynes in *The Economic Journal* in September and December 1940 respectively. In the latter of these, Madge asserts: 'Of the three, Coventry seems the freak town. For all its poverty, Blackburn shows a remarkable propensity to save; and for all its traditional prosperity, due to the balance and mixture of its industries, Bristol shows more analogies to Blackburn than to Coventry' (Madge 1940g: 410). The account of Blackburn draws heavily on Madge's experience of Bolton, as he delineates the cultural and historical components of this propensity to save: 'In Lancashire, saving institutions have a flying start. Lancashire is the cradle not only of the Co-operative Movement, but also of the Friendly Societies, the Holiday Clubs, the Mail-Order business, and more recently of a more speculative form of investment, the Football Pools' (430). However, it is clear that he used the same standards to judge Coventry: 'But to anyone with any experience in the economics of a working-class town, the crowds and the spending were not on a scale to equal the very high income levels attained by a large proportion of the population since the speed-up began (Madge 1940f: 327). The logic behind this rather peculiar stance was that, according to Madge, Coventry was a more predominantly working-class 'town' than Blackburn, with 87% so classified as opposed to 81% (Madge 1940g: 414). Therefore, he criticised Coventry for not behaving like Blackburn. It is this attitude that sets the general tenor of his account of Coventry, for although he notes that many inhabitants are new-incomers paying off past debts and sending money 'home', and that many are paying off mortgages, he remains implicitly judgmental: 'Just as Coventry has been during the past five years a paradise for the speculative builder, so it has been a happy hunting-ground for the hire-purchase merchant' (Madge 1940f: 339). Madge's overall argument – that the greater the surplus of income above the level of basic needs, the lesser the proportion of overall income saved (Madge 1940g: 420–1) – only makes sense if the underlying assumption is accepted: that the basic needs and cultural behaviour of the 'working class' can be standardised.

Keynes discussed Madge's findings with the Chancellor of the Exchequer, Sir Kingsley Wood, on 21 October 1940 (Keynes 1983: 821). Subsequently, Keynes arranged for the printer to supply twenty galley proofs of Madge's long article on Blackburn and Bristol, which included full comparisons with Coventry, so that he could circulate them to meet 'official interest' (822). By February 1941, the required decision had been made in the Treasury to raise £250 million in new taxation. During the six weeks to budget day, Keynes was involved in sorting out the finer details and drafting the budget speech, to which he added quotations from the Madge surveys. Moggridge records: 'On the day before the Budget [Keynes] worked 13 hours straight. Although he was rather knocked out the next day, he reported to his mother: "I went to the House of Commons for the Budget ... and thought the MPs in the mass a truly sub-human collection"' (cited in Moggridge 1992: 645–6). Even so, he was not too tired to write to Madge: 'I wish you could have been in the House of Commons this afternoon to hear the Chancellor of the Exchequer using some of the quotations from your Report about deduction of income tax at source. He had a big success – quite one of the bright moments of the afternoon' (Keynes 1941). The relevant section of Woods's Budget speech was as follows:

> I should like at this point to refer briefly to the scheme for deducting Income Tax from salaries and wages which was introduced last year .... In recent weeks an unofficial inquiry was made in a certain area as to how this new scheme was working. Four-fifths of those concerned definitely favoured the new method. I should like to quote some of the actual comments made on the new system, some from the men and some from the wives: 'Well, you are sort of paying for the war as you go'; 'A very good idea'; 'What you don't have you don't miss'; 'He doesn't mind it'; 'He did mind at first, but he has got used to it'; 'It isn't what you feel, but what you have got to do.' This shows how the Englishman has a genius for co-operating with the tax collector. In fact a first-class revolution in our fiscal system has happened, though silently, in the past year (Wood 1941: 1301–2).

This, then, was the point at which inter-war class instability was codified into the tax structure that was to fund the Welfare State and maintain the relatively rigid class stratification of the post-war British State. The working-class culture universalised in this process was that of the Lancashire industrial districts – a culture that Madge's own account acknowledges to have been defensive:

> Saving among poor people is an induced habit, not a natural tendency. In psychological jargon, it is the function of the super-ego to make people save – not many are natural misers. Therefore, because of the known weakness of the super-ego in face of temptation, Lancashire women mill-workers are prepared to pay somebody to come round once a week and collect their sixpences or shillings.
>
> ....
>
> We know that during the period of acute depression from 1921 to 1938 the Savings Bank has shown consistent increases year after year ... Why did Blackburnians stint themselves through these hard times? Partly because the uncertainties of the weaving industry led to great variations in family income, and families had to school themselves to save when there was money to save. Partly, I suspect, out of a half-conscious wish to get out of Blackburn (Madge 1940g: 429–32).

From Madge's April 1941 article 'Public Opinion and Paying for the War', it is clear that he regarded this condition of 'half-conscious' desire as potentially a form of collective neurosis: 'Behind Hitler, behind Stalin, behind Roosevelt is the will of the great industrial masses to win improved standards, whatever the political formula, however the formula is exploited by power-politics, and at whoever's expense the end is to be attained' (Madge 1941a: 36). The article concludes:

> Politically, I believe the formulation of post-war policy at home is far more important than the formulation of international 'war aims' so far as the masses are concerned ... the working class will accept any *equal* reduction of consumption, as they accepted military conscription. They will accept these measures just as long as the war seems worth while. The bad point will come when a substantial proportion are saying: 'We'll have to go on slaving the same, whoever wins' (46).

It is difficult not to see this as fear, not fear of the working class but fear of what will happen if the working class cross the line and thus blur the distinction between being-for-others and being-for-self. Madge quotes from a selection of working-class opinions to demonstrate political dissatisfaction, including: 'There'll be a revolution on the part of youth. I'm not a Socialist or a Communist, but I think there'll be something, and it won't be Fascism or Nazism, but something new. There's something that'll come forward, I don't know what it'll be' (45). Madge's concern over this indeterminacy can be linked back to the concern in *Britain* about an isolated 'home' falling prey to 'new leadership forms'. From this perspective, Madge's work for Keynes represents a rejection of the position advocated in *War Begins*: the establishment of a new compulsory unity. It appears that the potential within this position for the unchecked development of an authoritarian populism was personified for Madge by the example of Harrisson. This prompted Madge to reconsider the problem of 'home' – a process reflected in his study of Blackburn. The key sentence comes from a trade union official: 'Lancashire labour has long been immobile ... The substance of their life was home' (cited in Madge 1940g: 431–2). If home could be made secure both financially and emotionally, then the cause of working-class political dissatisfaction would be alleviated. Madge's project became the reinsertion of home into a political and public framework, which would protect it from the totalising tendencies of the media that had been identified in *May the Twelfth* and *Britain*. In 1941 it was necessary to begin by offering a positive point of identification to the working class: 'To be specific, I don't think that money is the sole driving force of working-class activity ... Prestige and social status are sought, even at the expense of economic needs' (Madge 1941a: 37). This prestige could only come from an equal status in society, an end to the outdated division of the 'literate and illiterate'. The distinctive M-O technique of direct quotation enabled the first stage of this process by allowing working-class voices from the street to be heard at the heart of the British State, in the House of Commons on Budget day. This historical moment is not usually associated with M-O because Madge had formally left the organisation by then: 'I have been asked to explain that since July 1940 I have no longer shared the responsibility for 'Mass-Observation...'' (Madge 1940g: 410n). However, it is apparent in retrospect that during the most crucial period of the war there had been two different versions of M-O in existence.

7
The Demobilisation of Everyday Life

Politics, Penguins, Pubs

The publication of Angus Calder's monumental *The People's War* in 1969 initiated a paradigmatic view of the home front in wartime Britain as a revolutionary 'ferment of participatory democracy' (18), which despite his own partial recantation and other revisionist attacks, still retains a powerful hold over the social imagination. It is a complex question as to why M-O is included in this configuration, despite the fact that both versions of it were working for a state bureaucracy which gradually gained control over the social sphere and dulled the promise of a new world. Obviously, its inclusion can be explained in part by the process already described in which the voices recorded came to signify more than Madge and Harrisson intended. Yet the extent to which these voices, either dissenting and assenting, were heard at the centre of government in the war cabinet and the House of Commons – and can still be heard by us today via publications stemming from the archive – was due to the drive and ambition of M-O's two leaders. Moreover, there is a strong sense that Calder's vision deliberately combines Madge and Harrisson's utopian dream of active leaders speaking for the led with the oppositional idea of the observers and observed giving voice to an autonomous emergent mass society:

> With parliament muted, with the traditional system of local government patently inadequate, with the army conceding the soldier's right to reason why, with the traditional basis of industrial discipline swept away by full employment, the people increasingly led itself. Its nameless leaders in the bombed streets, on the factory

floor, in the Home Guard drill hall, asserted a new and radical popular spirit. The air raid warden and the shop steward were men of destiny, for without their ungrudging support for the war it might be lost; morale was in danger.

'Morale' – that word which haunted the politicians, the civil servants and the generals. What the people now demanded, they must now be given. Had they taken the tubes as deep shelters? Oh well they must keep them ... Were they depressed by their conviction that victory would be the prelude to a new slump? Then plans must be made to ensure that life really would be better for them after the war.

So the people surged forward to fight their own war, forcing their masters into retreat, rejecting their nominal leaders and representatives and paying homage to leaders almost of their own imagination – to Churchill, to Cripps, to Beveridge, to Archbishop Temple and to Uncle Joe Stalin (18).

While, from Madge's perspective, Harrisson might have seemed to be going down the drain, and his ideas about leadership alternatively silly or sinister, it is clear that he remained committed to bridging the gap between leaders and led in pursuit of an elusive unity which always seemed to hang tantalising just beyond reach. This can be seen in his various struggles to form the 'wideawake unit' he had called for in the pages of the *New Statesman*, beginning with one of Adams's last attempts to use HI as a vehicle for their shared beliefs. On 6 March 1941, she wrote to Stephen Spender:

Dear Stephen,
You may remember that we spoke about the possibility of getting together a group of young writers to give us their observations and interpretations of morale at home. I'd very much like to discuss this further with you. Could you suggest the names of a small group whom we could invite informally? (Adams 1941b).

Spender's reply, from the *Horizon* office, suggested Louis MacNeice, Cecil Day-Lewis, William Empson ('Highbrow and obscure, but also pub-crawly and Mass-observant'), Arthur Calder-Marshall, John Sommerfield and Alun Lewis (Spender 1941). This configuration next appeared in public six months later when *Horizon* published the manifesto 'Why Not War Writers?' Spender, Calder-Marshall, and Lewis were among the eight signatories, along with Cyril Connolly,

Bonamy Dobrée, Arthur Koestler, Orwell and Harrisson. The manifesto advanced two principles:

1) *Creative writers must receive the same facilities as journalists.*
2) *Creative writers should be used to interpret the war world so that cultural unity is re-established and war effort emotionally co-ordinated.*
(Calder-Marshall et al 2001: 45–6)

Singling out the 'everyday lives of people' as an area particularly suited to creative writers, the manifesto asked: 'Why are there no novels of value about the building of shadow factories, the planning of wartime services, the operation of, shall we say, an evacuation scheme? Why are there no satires on hoarders, or the black market?' The four proposals at the end included calls for an accredited and financially supported 'official group of war writers' and for the international exchange of writers (46). The existence of drafts at the M-O archive confirms, which is in any case apparent, that Harrisson was involved in writing this manifesto. In a version dated 10 September, there is a fifth proposal at the end of the manifesto: 'The PEN club should be the primary organisation concerned in appointing and co-ordinating war writers, under official approval.' This is crossed out and the following is handwritten alongside: 'PEN confers to appoint a committee to report to Mr Brendan Bracken [Minister of Information] for the use of creative writers during the war and this to link up with J.B. Priestley's existing committee of writers' (FR867: 4). Priestley's 1941 Committee included Gollancz, Wells, Hopkinson, Kingsley Martin, editor of the *New Statesman*, David Astor of the *Observer* and Richard Acland, subsequent leader of the Common Wealth Party (Addison 1977: 188–9, 158–9). As Addison notes, 'The 1941 Committee ... [represented] a perfect snapshot of the new progressive establishment rising from the waves...' (189). It was a grouping of exactly the political, but not party-political, intelligentsia that Harrisson admired and Orwell loathed, which is possibly why the proposal disappeared from the printed version of the manifesto.

However, Harrisson and Orwell were both interested in the emergence of one of the 'imaginary' wartime leaders, Stafford Cripps. On 23 January 1942, Cripps had returned to Britain after having been the Ambassador to the Soviet Union. As Addison records: 'The 1941 Committee ... deputed David Astor and Tom Harrisson, as the young men of the future, to go and represent its views to [Cripps]' (198). At this point of the war, no longer regularly employed by the MOI or Naval Intelligence, Harrisson was supplementing M-O's freelance work by writing a radio column for Astor at the *Observer* (see Heimann 1998: 166–9). Later that year, this connection would come to Orwell's mind when pondering whether the revolutionary fervour being drummed up by the *Observer* was simply a plot by the Astors to get rid of Churchill:

> Mentioned this to Tom Harrisson, who has better opportunities of judging than I have. He considers it has a base in reality. He says the Astors, especially Lady A, are exceedingly intelligent in their way and realise that all they consider worth having will be lost if we don't make a compromise peace. They are, of course, anti-Russian, and therefore necessarily anti-Churchill. At one time they were actually scheming to make Trenchard Prime Minister. The man who would be ideal for their purpose would be Lloyd George, 'if he could walk'. I agree here, but was somewhat surprised to find Harrisson saying it – would have rather expected him to be pro-Lloyd George (Orwell 2001a: 340n).

Obviously there were limits as to who could be an 'imaginary' leader, but it is interesting to note Orwell's initially low estimation of Harrisson. Shortly afterwards, with no official connections left to pull strings, Harrisson was conscripted, but it was to be another two years before he flew out to Borneo to organise resistance behind the Japanese lines.

The fourth proposal of 'Why Not War Writers?' had been a demand that 'a proper proportion of these writers to be of groups most actively engaged in this war.' The true significance of this was revealed two months later when Harrisson's study of 'War Books' appeared in the December 1941 edition of *Horizon*. Over the previous two years, he claimed to have read 'literally every book which had anything to do with the war, reportage, fiction or fantasy':

> I have read *Secret Weapons* and *Attack Alarm*. I have seen how *Cheerfulness Breaks In* (via Angela Thirkwell), why *The Wounded don't Cry* (Q. Reynolds) and *Men do not Weep* (Beverley Nichols's latest and greatest horror). I have staggered through the anti-Semitism of Douglas Reed's *A Prophet at Home*, and sailed through the pleasant, easy documentation of Leo Walmsley's *Fishermen at War*. I have been borne through the air on *Bombers' Moon*; *Flying Wild*; *A Flying Visit*; *Winged Love*; *Air Force Girl*; *Wellington Wendy*; *Fighter Command*; *Fighter Pilot*; *Readiness at Dawn*; *RAF Occasions*; *Mysterious Air Ace*; *Shadows of Wings*; *War News had Wings*; *Winged Words*; *Wings of*

Victory; War in the Air; So Few. Never have I felt that I owed so little to so many (417).

He claimed that it was possible to identify a series of trends that had developed in succession across time:

1) 'The war in the country and the *evacuation novels*', most of which showed the evacuee in a poor light: 'very distinctly middle-class and often distinctly unsympathetic to the "masses".'
2) Diaries, notebooks.
3) Dunkirk books: 'Dunkirk at last gave people something to write about, and they wrote.'
4) RAF books.
5) Blitz books, which started when the heavy raids began: 'They completely reverse the earlier process, for now the working-class are one hundred per cent heroes. Extravagant admiration is lavished without regard for modesty, dignity or accuracy' (418–19).

Parallel to this transition from middle-class to working-class heroes, was the political shift to the right:

> War always gives a better opportunity to the narrow-minded and the intolerant, the tired man ... War makes it respectable to say jingoistic things about our own minorities, mean things about other majorities.
> .... To judge from most war books, Britain is fighting this war to protect the world against Auden and Picasso, the Jews and any form of collectivism (420).

As Harrisson notes, the left-wing slant of the 1930s had been washed out almost over night by uncertainty, opening the way to a return of traditional and orthodox themes. The key reason for this was easy to find: 'There have been a number of books by officers in the RAF and the Army, but not one by the far larger number of people in the ranks. There are many books about evacuees or blitzes, but none by working people who have been evacuated or blitzed' (421–2). The thrust of his argument, therefore, is that while the left intelligentsia of the 1930s were fighting or working for the state, the older right generation were shaping the representation of the war. So that although there was progressive publishing during the war, much of it by Penguin, it was outweighed by the weight of material, especially popular genre fiction (see Harrisson 1942a: 175–80), with a right-wing slant. Harrisson's analysis is particularly astute in the way that he links the political shift to the right with the parallel shift of focus from middle-class to working-class heroes. This connection seems counter-intuitive to those whose values were formed during the heyday of postwar Britain but it is important to remember that while the political changes to Britain during the war – the founding of the Welfare State – might have been *for* the working class, and while the working class themselves were portrayed as heroes, at the same time the dominant wartime representation *of* the working class, as inimitably summarised by Harrisson, was reactionary:

> The simple working man, usually the Cockney, and in nine cases out of ten either a char lady or a taxi driver. This character usually speaks for the unshakeable people of Britain, untainted by Communism, and for that matter untainted by anything else, except a pint of beer or an occasional bomb story in which the Cockney invariably shows heroic stoic qualities (421).

In the manner in which Empson argued that the 'Worker' is just another instance of the mythical cult figure, Harrisson understood this wartime representation of the working class as a means of providing a point of public identification with a revised symbolic order that served to preserve tradition against the possibility of the wider social transformation which had been anticipated in the late 1930s by people such as himself and Orwell.

However, the process was also more complex than Harrisson's account allows. For example, consider the celebrated wartime documentary *London Can Take It* (a.k.a. *Britain Can Take It*) made in the first weeks of the Blitz and immediately flown and shown in the United States, where its success countered a potentially disastrous perception that Britain was on the verge of collapse (see Jackson 2004: 231–2). The film was directed in two parts, the first by Harry Watt and the second by Jennings in his distinctive imagistic style. Watt, who was representative of that part of the Documentary movement which found Jennings too intellectual and condescending, was thoroughly appreciative of his contribution in this case: 'The famous last shot of a little Cockney workingman lighting a fag was one of Humphrey Jennings's touches of genius' (cited in Sussex 1975: 126). Taken on its own, this image can virtually be seen as the archetype of the kind of representation of the working class which Harrisson was to criticise. Yet in the context of the film as a whole, and the first half in particular, which shows people obediently trooping into the air-raid shelters as though part of the civilian army – a metaphor employed in the narration – advocated in M-O's *War Begins*, the closing image acts as a form of resistance to a dehumanising compulsory unity. Whereas the prewar *Spare Time* had combined collective activity and individual agency into a free identity, *London Can Take It* combined compulsory activity and class identity into a national unity that was itself resistant to Fascist aggression. As such, it was on one level entirely in keeping with the model of national unity advanced by M-O in *Britain*, which was also dependent on universalising a certain representation of the working class. Yet the difference between the two halves of the film also highlights the way that the equal weighting of the original model was unbalanced by the wartime shift in emphasis on to the primacy of showing unity in terms of the people's willingness to be led by their leaders. The Lambeth Walk had offered its participants a theatrical agency; *London Can Take It* offered a resistant gestural identity.

Of course, the development of this universalised imagery of working-class resistance, which from Harrisson's point of view was reactionary because it hindered the necessary structural transformation of society, was also a consequence of the actual conditions of the Blitz and as such probably a necessary development for Britain's ultimate survival. However, as the example of *London Can Take It* bears out, the exact shape of this imagery was conditioned by the pressure exerted by thinking like Harrisson's which, in attempting to overcome this resistance by insisting that change be imposed from above, highlighted and reinforced it. Under this tension, the cultural politics practised in the name of a classless society by those such as M-O, Empson, Common and Orwell broke apart. The trend of bringing the politicised sections of the new middle and working classes together by showing everyday life as a site of public contestation was reversed as workers were relegated once more to the resistant sphere of the everyday. In the case of M-O, the effect of this tension can be traced first through the initial split in *War Begins*, which left the founders at odds with the voices they were recording, and then on through the split between Harrisson and Madge. The general effect on society was the reintroduction of firm 'working class' and 'middle class' stereotypes, which had been rapidly losing their rigidity in the late 1930s.

To see the effect of this, we need only look at the 1947 M-O report on 'Penguin World', which was commissioned by Penguin (see M-O 1947a–c). One strand of this report was concerned with identifying a

'Penguin Public', which comprised those who read Penguins and consciously identified with the brand. Of the 50% of the population identified as regular readers (a third of the population never read books at all), 9% were identified as the 'Penguin Public', that is to say 4.5% of the general population. This group was far and away the most likely to have read a book during the previous twenty-four hours and read more in general, and bought more books of all kinds than other readers, frequently assembling their own personal libraries. M-O noted that 'the Penguin reader is likely to incline to the left politically' (M-O 1987b: 40). Whereas working-class Penguin readers behaved no differently politically to the wider working class, middle-class Penguin readers were five times more likely to vote Labour than Conservative, in contrast to the wider middle class, who were twice as likely to vote Conservative as Labour. Therefore, one can say that Penguin reading, while spread across the classes, took on a specific significance within the middle class, identifying a particular metropolitan fraction still identifiable to this day, although no longer by their buying of Penguins.

We can get an idea of the how this fraction conceived of themselves in 1947 from what M-O had to say about *Penguin New Writing*. This was a journal that was launched in 1940 by Penguin under the editorship of John Lehmann. It had started by reprinting articles from Lehmann's *New Writing* volumes of the 1930s and had been very much a continuation of 1930s concerns into the war with the first four issues including contributions from Auden, Isherwood, Spender, Orwell, Sommerfield, Gascoyne and B.L. Coombes amongst others. Published quarterly during the war, it reached peak sales of between 75,000 and 100,000 an issue (L. Jones 1985: 36–7). By 1947, it was down to only 2,305 subscriptions and in decline. The reason for the loss of *Penguin New Writing's* mass audience was illustrated by the account of a former constant reader:

Somehow the whole character of [the first issue] seemed to express the way I felt about things myself, and the way I should have liked to have been able to write about them. I probably had some sort of feeling too, that it might even be possible to write something myself for it ... I suppose I bought them quite regularly for three years ... Then I didn't get them so often ... I haven't bought one for over a year now, and I don't feel much inclination to do so. I know almost certainly what will be in them. I'm quite tired of bits of reportage which are supposed to be short stories – the stories about

working-class people which are apparently thought to be more real-
istic than stories about other people ... Anyway its not <u>New</u> any
longer for me (cited in M-O 1947b: 156).

Similar attitudes concerning the universalisation of working-class
culture characterise the M-O diarists of the same period collected in
Simon Garfield's *Our Hidden Lives*. For example: 'Middle-class dissatisfaction with the Government seems to be increasing ... The Government seems to be doing everything for the working class and
completely ignoring the fact that other classes do exist and have as
much right to existence' (cited in Garfield 2004: 375). By the 1950s,
the classless society promised in the 1930s had become a forgotten
dream.

One consequence of this cultural shift for the reception of both M-O
and the wider constellation of 1930s politics they exemplified, can be
seen most clearly in the change of attitudes to pubs. These had the particular function in the 1930s of offering a utopian space outside everyday class relations, as testified to by Common:

When two men of completely different class-origin have each tried
to do their best with the gifts God gave them, and each because
they wouldn't muck about with their talents, find themselves on
the wrong side of the door, then they can meet and share a PINT
together without any strained condescension, or theoretical
politesse. And that's the standard situation for the meeting of the
classes during the rest of the pre-revolutionary period through
which we take our uncertain steps. As it is so, let's have no weeps
about the tragedy, failure and frustration. It isn't frustration when
good men meet and like one another, even though they had to
throw aside their accomplishments in order to do it. It's a celebration, just a quiet one, held on the nod you might say, but a
rehearsal for the big splash (Common 1988: 47–8).

The beginning of Madge's poem 'Drinking in Bolton', first published
in the new series of *New Writing* which appeared in Autumn 1938,
similarly celebrates this aspect of the pub:

Not from imagination I am drawing
This landscape, (Lancs), this plate of tripe and onions,
But, like the Nag's Head barmaid, I am drawing
(Towards imagination) gills of mild,

The industrial drink, in which my dreams and theirs
Find common ground. (Madge 1938e: 46).

However, as has been suggested (see Milne 2001: 69), the ending of the
poem seems to imply the impossibility of this imagined community:

And in this hour are crowded all men's lives,
For, as they drink, they drown. So final night
Falls, like a pack of cards, each one of which
Is fate, the film star and the penny pool.
You sit there waiting for the spell to break.

It could further be argued that the poem 'expects and confirms' the
loss of working-class cohesion in the face of commercialised leisure (see
J. Roberts 1998: 63). Yet it is also possible to argue that the poem
simply predicts the imminent collapse of what at that point of time in
Bolton was still a mutually shared consciousness, suggesting neither
impossibility nor complicity. Indeed, the last line is reminiscent of
Madge's comment on surrealism that the spell is only complete at
certain lucky moments (see above: 87), which suggests that perhaps he
never expected that moment to last or thought of it as anything other
than a lucky coincidence. This would not correspond with a denial of
the potential of technology in the service of modernist technique to
transform society, so much as with a recognition that the ownership of
the means of technological production remained concentrated among
the sections of society least likely to promote such transformation
disinterestedly.

It was already apparent long before the Autumn of 1938 that M-O
would always struggle both to fund itself and to exert influence
because it could not compete with the media in terms of scale.
The only solution to these problems was to become employed by the
state which, as we have seen, pulled them further away from the imagined community and turned the memory of the pub into something of
an oasis. This was a sense that was acknowledged by Basil Nicholson's
'Introductory Note' to *The Pub and the People* when it eventually
appeared in 1943: 'You walk back into a warm bright room and marvel
that in 1938 we never knew that those spittoons were in Arcadia' (xii).
Despite the fact that the book's conclusion clearly notes that 'it must
be emphasised that this volume is isolated from its context' (350), it
was this nostalgic sense, which was not present in 1937 when the book
was researched and originally drafted but a product of the wartime

conditions under which it was published, that prompted a contemporary reception of the book as confirming the loss of working-class cohesion in the face of commercialised leisure: 'the whole trend of the age is away from creative communal amusements and towards solitary mechanical ones' (Orwell 2001b: 321). A nostalgic idea of the pub became fixed in the social imagination as another example of everyday resistance to the transformed but authoritarian wartime society. Consequently, as the return by Harrisson, Sommerfield and some of the others to Bolton in 1960 discovered, 'pub behaviour' remained as unchanged as 'everyday gesture', not just during the war but for decades after, while the rest of society changed away from it (Harrisson 1961: 42).

This uneven development has made it difficult for postwar critics to assess the pub encounters between Worktowners and observers described in *The Pub and the People* and the reports from which it was compiled. For example, Peter Gurney has used a report by Sommerfield to illustrate the co-mingling of bourgeois disgust and desire he sees arising from the encounters between middle-class observers and working-class female sexuality. Sommerfield's elision of '2 girls ... singing loudly a dirty song' into 'dirty girls' and references to 'a constant stream of women up to the lavatory' are taken to demonstrate his construction of working-class women as incapable of self-restraint. This impression is apparently reinforced by Sommerfield's accounts of the 'dirty girls' stroking his thigh and playing footsie. Gurney concludes: '[The two women] were out to have a good time in Blackpool, probably knew about Mass-Observation and found it entertaining to lead this stranger on and cast him in the role of innocent stooge, an appropriate role to their vampish yet comic performance. Sommerfield, for his part, read these signs literally and failed to appreciate what were very likely playful deconstructions of male seriousness and status' (Gurney 1997: 279). This is taken as illustrating how M-O, in the characteristic manner of anthropology, preserved the sovereignty of the masculine bourgeois subject, as represented by Sommerfield, against potential decentring by denying the performance of the 'dirty girls'. Yet the final sentence of Sommerfield's report, which is included in Gurney's account suggests the complete permeability of class and gender barriers: 'At 11.30 we all leave, arms linked ...' (cited in Gurney 1997: 278).

Quite clearly, Gurney's position is only sustained by denying Sommerfield the same ability to transgress his perceived social class that is freely accorded to the working-class women. This seems particularly ironic in Sommerfield's case because he was sufficiently adept at transgressing social class to be often described as a working-class writer (see Cunningham 1988: 306–8; Hayward 1997: 48). Indeed, his former M-O colleagues gave a variety of opinions as to his background encompassing the full range from Barefoot's assertion that he had been to Winchester, to Cawson's understanding that he was proletarian (Barefoot 1939: 70; Cawson 1980: 10). Madge's judgement was nicely measured: 'John Sommerfield was not particularly working class ... He was a man without any side and I would have thought he melted into Bolton life pretty easily' (Madge 1978a: 2).

The fact of the matter is that perceptions were different in the 1930s but the fate of the Worktown pub material in particular has been to be viewed through the lens of a rigid postwar class representation which was generated by wartime tensions. Therefore, the strongest point of M-O's work, their capturing for ever of a period of liberatory intersubjectivity, becomes portrayed as a voyeuristic and intrusive 'crossing of the line'. The great pity of this reception with respect to the pub book, is that it has obscured the political aims of the book, which were to defend exactly that sense of the pub as a public and a political space that became eroded in the postwar period. There is a section dealing with the history of the pub which locates it, in terms analogous to Habermas, as the site of a working-class public sphere in which newspapers could be read and trade unions could meet (see M-O 1987b: 82–4). Not only is the pub portrayed as spatially liberated but also as temporally liberated: 'Alcohol helps the worker's brain to escape from the speed at which it has to function during worktime. For a short while each week the pubgoer is physically and psychologically emancipated from the restrictions of normal Worktown life' (199). The consequent effect of the book is very similar to that of *May the Twelfth*, in that it holds open the prospect of a temporally and spatially liberated public sphere by deploying image-facts in a manner that contests the homogenisation and depoliticisation of everyday life. Or to be more precise, as previously discussed, the book complements *May the Twelfth* by reversing its focus of revealing the philosophical independence of the masses, in order to reveal the materiality equally required for mass independence. However, with the collapse of that emergent classless period, the imagery was left to reconstellate into the resistant everyday formations which have come to be the dominant cultural representations of both pubs and the working class.

Society in the Mind

In the *New Statesman* of 28 September 1946, Harrisson complained that the first thing he had noticed after two years in Borneo was the horror of having 'to stand solitary in a city pub, or face your fellows in a train speechless, almost ashamed' (Harrisson 1946a: 221). The intersubjective exchange which M-O had promoted by creating a form of public sphere was no longer immediately possible because people had shrunk back into 'the restricted circle of private western experience'. He summed up the overall situation in which everyday life had become merely a site for the theatrical staging of the everyday succinctly: 'In the same way as in Munich days, people are repressing the inevitably depressing ... The struggle to live is almost wholly domestic. The squatters were a wonderful relief (whatever our opinion about them), something outside ourselves which was *dramatically* domestic.'

Under such circumstances there was nothing to do but plunge back into M-O, which was now coming under organised attack from unexpected quarters. The first issue of *Pilot Papers: Social Essays and Documents*, a new journal edited by Madge, included an article by Chapman which, amongst other things, laid down the criteria for the representativeness of social surveys: 'an inquiry will only be representative if it is based on either a representative sample, random sample or a census' (Chapman 1946: 81). A few sentences later, a curt note adds: 'The Mass Observation inquiry into 'People's Homes' suffers from this same defect.' Harrisson was in correspondence with Chapman from the moment of his return to Britain, mostly about research Chapman was conducting into tenant morale, but he did refer to Chapman's article when mentioning a forthcoming one of his own:

> I hope that the article that is coming out in the January number of *Pilot Papers* will be accepted by you in the friendly and constructive spirit in which it is intended. At least I do not dismiss everything that you have to say in a one-and-a-half-line paragraph. Though I suppose I could do so by proving that your article, being written by you, is not 'representative' (Harrisson 1946b).

Harrisson's 'The Future of Sociology' – the title confers a retrospective poignancy because what he outlined was certainly not to become the future of sociology – is probably the most concise and powerful expression of his approach to social research. He does not discuss M-O directly, but the way he constructs sociology anticipates the responses of those later writings about M-O which have attempted to outline an avant-garde sociology (N. Stanley 1981; Chaney and Pickering 1986; Highmore 2002a). Therefore, it is quite fitting that it should echo Jennings's review of *Surrealism*:

> Sociology has one function as 'scientific reportage', so 'useful' to advertisers, politicians and planners, and so easily made presentable on a single type of statistical method, that this can the more easily cause a lop-sided development, an obsessive interest in only a fraction of the whole human field (16).

This is in the context of an argument that social psychology should concentrate on normal situations rather than abnormal ones or 'social problems'. Although Harrisson does not say so, this is what characterised both phases of the Worktown project as opposed to other social surveys of the late 1930s which focused on poverty and unemployment. He directly refers to Chapman's strictures on representativeness and asks:

> Representative of what? Take a single example. Random sample *questionnaires* have lately asked people about church and pub-going. The results give a decimal point picture of proportions affected. Yet they bear little relation to the actual number of persons who *do* go to pubs and churches – because many people think it socially respectable to tell a stranger they go to church, and some people think it not so respectable to admit pub-going (17).

Such an argument is commonplace today after repeated experience of misleading poll results with respect to political elections has led to a widespread distrust of the procedure. Similarly, he points out that a sample giving the same results when enlarged does not prove representativeness, but merely homogeneity. Particularly ironic with respect to M-O's reception history, are his opening comments that the obsession with statistics and quantitative research are an understandable reaction to the tradition of totalising sociological theory as represented by figures like Herbert Spencer. Subsequently, of course, the long term effect of the pendulum swinging back violently in the other direction at the beginning of the 1970s was that M-O would eventually become subject to theoretical critiques such as the Foucauldian arguments of Gurney. The gulf that Harrisson identifies between the extremes of 'philosophical, subjective sociology' and 'statistical, quantitative sociology' (10) is still very much in existence in Britain. Harrisson refers to an 'understanding sociology' (20) but there is no indigenous tradition of such an approach apart from one that would run through M-O, so it was always next to impossible for M-O to ever legitimate itself.

The existence of this article is evidence that M-O did try and legitimate themselves in the face of the Clapham Report's vision that social research should be government and university based. An alternative social science forum was set up at PEP, with Harrisson as one of the convenors (see Harrisson 1947b). However, Harrisson left Britain for Borneo at this point and passed on his role as convenor to Madge. In the end, it transpired that he never came back to take up the reins of M-O, which became a limited company in 1949 with Len England, an observer since 1939, becoming the Managing Director. Harrisson was careful to keep the rights to all the material from 1937 to 1949.

One of Harrisson's fellow convenors at the alternative social science forum was Mark Abrams of the London Press Exchange, an opinion polling company that was a direct competitor of M-O Ltd. In 1951, Abrams devoted a whole chapter of his book *Social Surveys and Social Action* to a critical attack on M-O which, perhaps more than any other account, conditioned its subsequent negative image in sociological circles (see L. Stanley 1990: 3–4; Sheridan et al 2000: 37). Reading it today, it is impossible to see why anyone would take Abrams's criticisms seriously, other than through vested interests, because his chapter consists almost entirely of sweeping and repetitive generalisations along the lines that M-O collected 'dreary trivia' (107) and 'boring and unrelated quotations' (108), as well as the accusation that their observers were untrained. In fact, it was very much a reprisal of the prewar criticisms of Marshall, Jahoda and Firth. Yet there is much less excuse because Abrams would have known of the work done by M-O during the war and that far from its personnel being untrained, their work for M-O was seen as good training for subsequent careers at the GSS, as evidenced by the careers of Box, Thomas and, later, Bob Willcock.

In Abrams's defence, it can be argued that the problem was that there was no theoretical tradition to relate its work to and that, furthermore, the very structural changes of society brought about by the war had, as we have seen, altered the context in which M-O was received: 'It is likely that the average man-in-the-street would regard the Mass-Observation volunteer as a bit of a crank' (111). But, it had been this feature of appealing to a self-selecting membership that had characterised M-O's initial success in identifying emergent trends in society as opposed to simply recording representative snapshots of the status quo. It was the social change that turned the new middle class from an emergent force in the late 1930s into an oppositional fraction in the late 1940s which ultimately did for M-O.

Madge demonstrated an understanding of this change in the conclusion to his supportive editorial introduction to Harrisson's article:

> There is much more tolerance in this paper by Tom Harrisson than in his writings of ten years ago, though there is still enough attack to make some people annoyed and uncomfortable. However let them tell themselves that without such intellectual gadflies we should never overcome our very strong resistances to direct social fieldwork – resistances less marked in America and related to English inhibitions, English shyness and sensitiveness about class, English love of privacy and dislike of 'confessional' techniques. If in spite of all these obstacles, quite a bit of useful social observation has been done in Britain, it is because of the empirical streak in the English character, the English love of facts and distrust of systems. The qualities which have made Englishmen good naturalists can also make them good sociologists, in time (4).

It is another argument for bringing in the absolute less prematurely. Like Harrisson, Madge had noted the paradoxical nature of the Communist backed Squatter's movement of 1946; where people were organising themselves politically and socially in the attempt to get permanent housing, but the type of housing that most of them desired to live in was a prefab, which in practice encouraged very little communal activity even with next door neighbours. Madge was confident that with 'a little encouragement', a natural balance could be restored: 'The Englishman's home is his castle *and* he likes to be one of a crowd' (Madge 1946: 6). This was what he hoped to achieve when, after the finances for *Pilot Papers* collapsed, he took a position as Social Development Officer at Stevenage, stating unequivocally, 'New Towns must be planned as carefully and scientifically as, shall we say, the invasion of Normandy' (Madge 1948: 110). This involved a return to a logic similar to that behind the theoretical framework of the three social areas. Madge envisaged a utilisation of space that would structure patterns of living into the linked concentric circles of 'indoor private space', 'outdoor private space' and 'outdoor public space'

(Madge 1950: 136) and so help to restore the depoliticised postwar private sphere to a rebuilt public sphere:

> My own personal view is that the whole geography of the housing estate needs turning back to front. Instead of the front-doors and the living-room windows opening on to the asphalt street, with a parallel row of houses returning stare for stare, I envisage a living-room looking on to what is now the back-garden. This would no longer be just a cabbage patch and yard for what are called utilities. It should be a private space, pleasant to walk out into, and leading to further spaces of diminishing privacy, opening out on to trees and grass, tennis courts, children's playgrounds and swings and paddling pools, all of which would be part of what we call the garden common (Madge 1949: 268).

However, the overwhelming majority of postwar housing was not like this and the New Towns, themselves, never fulfilled their promise. The financial constraints of the late 1940s prompted a move from social development as an integral part of the process – one of the factors in Madge losing his job – and also drove rents up to a point beyond many working class incomes (see Madge 1951: 246–7). After Madge was appointed to the Chair in Sociology at the University of Birmingham in 1950, he turned much of his attention to development and education issues in South East Asia.

Yet he by no means abandoned M-O as he acknowledged in the postscript he wrote for *Britain Revisited* at Harrisson's request. There is a certain knowing irony in 'Professor Charles Madge' following in the shoes of Malinowski as a legitimating expert for the unorthodox approach of M-O, but his analysis was acute, suggesting that M-O had always sought a more 'imaginative and active kind of sociology' and that Harrisson's comments on that matter in *Pilot Papers* remained as pertinent in 1961 as they had been in 1947. Madge acknowledges that M-O had political roots but notes how quickly priorities changed for him: 'Personally I soon found it impossible to combine political with sociological activity. Even as a socialist, I felt I must put sociology first' (279). He then develops this personal analysis into a general statement:

> In the thirties, poetry and politics and nascent sociology grew confusedly side by side. In the fifties, poetry and politics had lost, at any rate in Britain, the revolutionary impulse of the thirties. Sociology was slowly growing, in an institutional sense; but it was mostly

something to talk about rather than an act of exploration. The fifties were in the doldrums, intellectually and aesthetically. But from the point of view of 'ordinary people', in Britain and in many other countries too, it was a good decade. Industrial populations were better fed, better housed, better educated, better informed, had more money to spend, were more widely travelled than ever before. This was the 'change' that was most apparent in Britain revisited (279).

This can be read as a mental weighing up of what was lost and gained by Madge's own actions. In particular, it revolves around his decision in late 1937 to adopt a stance of public social scientific respectability and restrict his poetic and political impulses to a more personal level with respect to M-O's work, which is not to say that he did not publicly continue with other political and poetic activities – because he clearly did. Therefore, the depoliticisation of postwar society with its negative effect on not just politics but poetry as well – including Madge's failure to persuade Faber to publish his postwar poetry – had to be offset by the material betterment of ordinary people for which Madge, given his role in the 1941 Budget, could feel a degree of personal satisfaction.

Yet the persistence of what Harrisson called 'unchange', in everyday behaviour and the pubs, suggested an ongoing need for M-O. Madge reflects that viewed in retrospect it was the scope of early M-O, rather than its early ambitions, which was of true value then and which deserved to be reestablished in the sixties: 'Compared with stodgier, if steadier, sorts of social inquiry, it assumed that a wide range of human phenomena had serious significance, and it opposed a narrowing of the spectrum of scientific concern' (280). The need was for an early-style M-O to investigate the remaining pockets of irrationalism in the country and the conditions at the beginning of that new decade once again seemed to be favourable for such an enterprise.

The theoretical framework behind this idea is discussed in his 1964 book *Society in the Mind*, which returns to Bateson's concept of social eidos and attempts to map out the social and historical structure of its contemporary stage, in which 'along with a plentiful inheritance of mystery and morality, there is an overriding tendency to look on society as a working system, which can be made to work better and better by rational-technical means. Central to this conception are first, the idea of progress, second, the economics of the market and third, the politics of the ballot box' (55). Against this, Madge describes the role of inarticulate resistances – sources of the 'unchange' discovered in Bolton by the revisiting mass-observers of 1960 – such as 'the resistance to school' and non-participation in religion and politics (123–4). This is what he calls the un-social eidos:

> The consciousness of the social self is constructed with the help of social ideas, with elements, that is, from the available social eidos. Each individual works out his individual version of this social self. But it is arguable that at an equally realistic level, each individual could work out – and in fact normally does work out – a consciously un-social self, prepared to resist or evade those social demands which it may begin to consider on realistic grounds to be excessive (138).

He goes on to argue that 'the canons of mature un-social eidos ... have taken shape ... largely through the activities of poets, painters and novelists, whose influence has been felt not only through their artistic works and manifestos but through the example of their own private lives' (139). As Milne points out, this implies 'the idea of poetry ... as a private, almost anti-social resistance to dominant values' (Milne 2001: 73). Indeed, the argument appears very similar to the one in 'Poetry, Time and Place' concerning how Marvell's personal poetic independence underwrote his politically active public role which, as we have seen, can be taken as a model for Madge's own adoption of a position of public respectability as a social scientist.

Read together with the afterword to *Britain Revisited*, it is possible to argue that Madge's aim here was to reverse the emphasis on scientific respectability in the late 1930s, which had by the early 1960s achieved its aims of promoting material betterment, and focus instead on a more personal aesthetic sense. Obviously, the conditions seemed favourable to such considerations, but Madge's ambitions exceeded promoting alternative lifestyles and were directed towards 'the groundwork of a potential theory and programme for the socially controlled relaxation and liberalisation of the social bond' (140). Thus, he had in mind not so much a privileging of unsocial eidos as a realignment of social eidos brought about by the adoption of an aesthetic teleology (see 143, 146). By pointing out that 'art is in touch with a sphere of mental reality in which illusions are true' (152), he suggests once more the possibility, which he had moved away from at the end of the 1930s, of an aesthetic or poetic kind of thinking which the reality principle does not veto because it is powerful enough to change reality.

The fact that the argument is now stated in terms of Bateson's vocabulary, suggests a recognition that Bateson had identified a form of ritual which did not simply ensure the continuation of the symbolic order, but which enabled the conscious alteration of that order. The basis of this complex agency was schismogenesis, a dialectical process of fission and fusion capable of reaching a mutual balance across a web of networks between any combination of individuals, groups and societies: 'This process occurs not only between groups but also between pairs of individuals; and on theoretical grounds, we must expect that if the course of true love ever ran smooth, it would follow an exponential curve' (Bateson 1958: 197).

The Resumption of the People's War

As a process of change responding to the contestation of cultural norms, schismogenesis could as easily result in fission as fusion and, as Bateson had observed, in the fission of European societies 'the daughter groups generally separate from the parent in revolt against the elaborate centripetal hierarchies – legal, religious and military – which are a characteristic feature of the integration of these communities' (270). The cultural flux of the 1960s resulted in exactly such a case of fission. In the introduction to *The People's War*, Calder noted how 'current events in the western countries indicate that a younger generation is consciously resuming the quest for democratic community which was thwarted in the later years of the war' (19). In September 1968, Madge became Dean of his faculty with the intention of increasing student participation in its running. However, even as he implemented this, events were overtaken by the national surge of student radicalism. Madge was not unsympathetic to the resulting student occupation at the University but was also aware of what he termed the 'irrationality of "direct" political action' (Madge 1987: 247).

The stress of this period led him to opt for an unofficial early retirement, but he was still at the University, although no longer Dean, when what was to become known as the 'Atkinson Affair' arose. Madge's former 'Dean's relief', M.R. Atkinson had been strongly supportive of student activism. At the end of his tenure, he had left the university but he subsequently applied for a permanent post. The department selection panel voted narrowly in his favour but were subsequently overruled. The decision was discussed in the national press in terms of academic freedom and one of the consequences was that the British Sociological Association called on its members to boycott applying for posts at Birmingham until their selection procedures were clarified. Madge's successor as Dean had a nervous breakdown. As Madge commented in his autobiography: 'The whole series of events completed my disillusionment with Sociology in its current phrase, and with other forms of Marxism and Leftism' (249). A further moment of disillusion ensued in 1975, when Madge attempted to review *Pre-Capitalist Modes of Production* by Barry Hindess and Paul Hirst only to give up in horror on finding the concepts of history and social science that had sustained his postwar work rejected as valueless mystification, leading to the 'realisation of what a make-believe, really, all of my past activities had been' (259).

In the meantime, Harrisson, newly installed at the University of Sussex, found the heady atmosphere of the early 1970s a welcome return to the vibrancy of the 1930s and, not content with a passive role managing the archive, quickly resumed M-O activities such as taking a team of observers to the Bromsgrove byelection – where Labour were to gain the seat on a 10% swing that was seen as an expression of public hostility to the Common Market. Harrisson published the results of his investigation in the *New Statesman* and concluded: 'More than anything else in the M-O Archive I was reminded of the reactions after the retreat from Dunkirk. There was then an intense relief: relief that we had got back, more or less intact, *out* of Europe, safely behind the white cliffs of Dover' (Harrisson 1971: 798). He was forthright in denouncing the anti-foreign undertones of this 'Little England' as 'racialist'. It is possible to see Harrisson's work in the 1970s as a form of guerrilla war against the Dunkirk spirit. As he wrote at the time, there was much 'badness' in the war 'which it is deadly dangerous to glorify for the unborn' (Harrisson 1975: 339). His untimely death in a road accident the following year did little to halt this attack as can be seen from the preface to the posthumously published *Living Through the Blitz*:

It has proved something of an advantage to this writer, co-ordinating and necessarily *selecting from* a mass of old records, that he had an unusually wide experience of living through the blitz. It has been a greater advantage, however, that he has not been subject to the subsequent three decades of brain-washing.

.... In the past some critics have tried to discount the whole of M-O as leftist or dilettante. The records of the men and women, active in the work both then and since, refute this fallacy. If any such charge is repeated in the fourth quarter of the twentieth century, it may only be because some Britons, especially responsible ones, cannot face the full facts about their 'finest hours'.

.... At no time in World War II generally and in the blitz particularly were British civilians united on anything, though they might be ready to appear so in public on certain issues (13–15).

Although the book discussed the above-mentioned 'full facts' in detail, such as 'trekking' – the daily refugee cycle of mass-migration from target cities to seek shelter in the countryside (165–8) – its effect came not so much from revelation as from its tone, which debunked both official and popular versions of the war. Therefore, it helped pave the way for a subsequent wave of revisionist history led by Angus Calder's *The Myth of the Blitz*, in which 'myth' was used in the sense defined by Roland Barthes and the 'blitz' as a referent to a whole series of symbolically linked events from the period 1940–1941, including the evacuation of the British Expeditionary Force from Dunkirk and the aerial Battle of Britain as well as the actual German bombing raids (1–3). Calder argued that these events have never been seen as part of the wider European and World history of the time, because Britain was never invaded and they happened before the entry of the USSR and USA into the war. They 'have acquired [an] ... aura of absoluteness, uniqueness, definitiveness ... these were events in which the hand of destiny was seen' (1). In particular, Calder took care to show how his former construction of the People's War had become complicit with the Myth (xiii) and to examine how the Left had not captured history in 1940 so much as to allow history to capture them (15). However as the Myth was above all a myth of British unity which came to encompass the working-class majority of the population, the Left quickly came to have the biggest stake in it. The net effect remained 'a juster and friendlier society' despite the fact that the Myth had subsumed the more radical transformatory dynamic of the People's War: 'If a disastrous conflation of state with community produced an excessively bureaucratic welfare state out of control by the People whom it professed to serve, at least the Myth had fostered the notion of the mutual responsibility of all for the welfare of all'(272).

This idea of a poised tension between the People's War and the Myth neatly captures the sense of M-O's position in the war, caught as we have seen in a situation where the more they pressed for accelerating the pace of social change, the more their own collection techniques registered everyday resistance. While the models of representation that M-O pioneered in the late 1930s, especially in *Britain*, and which were continued not just by them but by the whole constellation around the 'Social Eye of *Picture Post*', were able to incorporate these everyday images into a unified narrative of Britishness; more pressure from the top of the organisation could change the direction of that narrative. Thus the internal dynamics of M-O serve as an example of the wider historical process. The advent of the archive opened up a new dimension because it suddenly made all of those everyday images of the war available as a primary source for incorporation into new narratives of wartime Britishness, and by extension Britishness *per se*, designed to respond to new contexts in a manner exemplified by either of Calder's war books. The launch of the new project in 1981 established M-O as an ongoing project of 'Writing Ourselves and Writing Britain' and confirmed that, as an astute *New Statesman* reporter concluded in 1987, 'the people's war continues, based in Sussex' (Khan 1987: 26).

The origins of the new project lay in the decision of Philip Ziegler, who was using material from the archive to write *Crown and People*, for publication in the queen's silver jubilee year of 1977, to collect new material through friends and former mass-observers. This inspired David Pocock, the Professor of Social Anthropology who had become Director of the Archive following Harrisson's death, to launch the 'Mass-Observation in the 1980s' Project with a directive addressing a number of issues including the forthcoming 'Royal Wedding' between Prince Charles and Lady Diana Spencer. This was quickly followed by the announcement of a day-survey for the wedding, reinforcing the extent to which the birth of the new project replicated the original M-O's concern with royal events (see Sheridan et al 2000: 44). Indeed, the concern of reincorporating the founding project into ongoing studies has a long history within M-O.

The chapter on 'Royal Occasions' in *Britain Revisited* pairs extracts from *May the Twelfth* with reports from the 1953 coronation to show the similarity of both in terms of 'official and unofficial behaviour' (Harrisson 1961: 233). Survey results indicate that in both 1956 and 1960, 90% of the population supported the monarchy in preference to a republic (231), leading eventually to the conclusion: 'it seems unlikely that existing attitudes to royalty will be drastically or lastingly changed' (251). It was very much in this tradition that Ziegler used material from the M-O A, and *May the Twelfth* in particular, for his book studying the attitude of 'the man in the street' to the British monarchy (Ziegler 1978: 9–13, 212–13). The M-O material serves to demonstrate the undoubted widespread participation in the coronation activities nationally, and to comment on the 'commonly encountered' patriotism, 'sometimes almost unwillingly expressed...' (11). Ziegler invokes the 90% pro-monarchy statistic in order to state, that in the event of a referendum on becoming a republic, no one could 'seriously doubt that those weeks would be dominated by the thunderous roaring of the royalists ... and that ... Britain's electorate would prove by their votes what they had demonstrated by their cheers in 1937, 1953 and 1977' (197). The book concludes that 'the British want the royal family, that their reasons for doing so are sensible, even meritorious, and that our national life would be impoverished if the monarchy were to be eliminated.' (203).

This trend was continued by the 1987 republication of *May the Twelfth* with an afterword by Pocock, who expresses pleasurable surprise at the 'apparently identical features in the reactions on that Coronation Day and those reported in the Day Diaries written for M-O on 29 July 1981, the wedding day of Their Royal Highnesses, the Prince and Princess of Wales' (Pocock 1987: 420). We learn of the young man who 'sedulously and even ostentatiously avoided anything to do with the Royal Wedding' until he heard something about it on the BBC World Service and ' "was unexpectedly moved ...Felt absurdly, nicely, English"' (cited 421). This is contrasted with a report by a typist in *May the Twelfth*. According to Pocock, 'Her feelings also changed as she was seduced by the public celebration and if , at thirty-nine, she was more suspicious of these awakened emotions, it is not surprising.' In fact, her 'suspicions' had been concisely stated in political terms: 'It [the pageantry of the coronation] is too dangerous a weapon to be in the hands of the people at present in power in this country' (Jennings and Madge 1937b: 305). The back of this paperback reissue announces 'that in a half-century that has witnessed dramatic changes in our daily lives, underlying attitudes have changed remarkably little.' Pocock concludes by analysing why the same reactions are always present:

Kingship is an institution much older and more complex than constitutional monarchy and more primitive, in the sense that it is an expression of a powerful human need. For evidence of this we have only to contrast the rational demands of constitutional monarchy with the popular insistence on royalty. There is no constitutional requirement for the monarch's family, other than her immediate successor perhaps, to play any part, let alone be invested with the distinctive glamour of royalty; no presidential figure could evoke the insatiable curiosity about its private life that the Queen evokes – all this is the creation of the public, it is what we wish to be so.

Sometimes one reads the comment that the British Royal family is to be likened to one or the other of the unending television serials about some wealthy family. The judgement could not be more superficial: it is rather the reality of royalty and the distinctive complex of emotions which royalty alone can evoke, that accounts for the popularity of coloured shadows (423).

Whereas the original publication of *May the Twelfth* reveals how the media diluted the formerly unique relationship between the king and the people to just one moment in a chain of differences, Pocock's argument tries to show that the presence of many media moments highlights the uniqueness of the relationship between king and people. Thus where anthropology had been used to show the reality of social change, social change was now being used to demonstrate the reality of anthropology.

Therefore, it is hardly surprising that Tom Nairn has taken *Crown and People* and *May the Twelfth* as demonstrating M-O's complicity in constructing what he has called the People's Monarchy, a configuration in which the royal family's status as neither 'Them' nor 'Us' serves to embody a collectively unconscious Britishness (Nairn 1998: 22). The 'glamour' of this monarchy operates as 'an interface between two worlds, the mundane one and some vaster national-spiritual sphere associated with mass adulation, the past, the State and familial morality, as well as with Fleet Street larks and comforting daydreams' (27). Yet, Nairn argues, the reality of this glamour is no more than 'our collective image in the mirror of the State' (13). This is true but it underestimates the scale of the reciprocity of this collective image. *May the Twelfth* tried to liberate this collective image in an unequal battle against the media. *Britain* combined what had been learnt from that project with what had been learnt from the Worktown project, and created a fairy-tale resolution by way of *Me and My Girl* and the Lambeth Walk, which allowed the working-class majority to embody national identity in an inverse relationship with the monarchy that became part of the wartime myth of Britishness. The saving virtue of the relationship lay in its anthropological aspect: that the antiquity of the relationship between crown and people bound both parties equally to a traditional symbolic order which therefore was not subject to the same kind of manipulation as the purely media relations which enabled the structurally similar arrangement of Fascism.

Therefore, just as it has remained connected to the linked paradigms of People's War and Myth of the Blitz, M-O remains connected to the linked paradigms of monarchy and media. The dominant collective self-definition of observers participating in the new project is that they are 'ordinary'. This ordinariness is most typically constructed as a supplement to the media. That is to say, people write for the current project because they feel that ordinary voices are not represented in the media (see J. Thomas 2002: 38; Sheridan et al 2000: 214–19). However, this ordinariness could also be seen as a supplement to the monarchy in the same way as the relationship was constructed by Ziegler and Pocock. The nature of an archive is that it potentially makes those ordinary voices always available for that kind of appropriation, regardless of the fact that the current organisation would not actively support such uses. This is why M-O publications and their original emphasis on presentation have to be considered as central to the M-O legacy as the archive material.

The importance of this is upheld by one of the most recent publications to be based on archive material: James Thomas's *Diana's Mourning: A People's History*. Although the title suggests another addition to the annals of the People's Monarchy, this particular organisation of ordinary voices transcends mere ordinariness, to show how the events around the death and funeral of Diana did provide, however temporarily, a public space in which ideas and values were contested. As Thomas observes: 'Diana's mourning far from serving as a basis for a new national identity, marked a serious challenge for most people to the idea of "Britishness"' (3). He cites the concerns of one observer: 'I hate to think how the whole "Diana death" business will become, has already become, a homogeneous myth, like the 'chirpy cockney in the blitz' myth along the lines of "a nation mourns"' (7). This sets the book up to use observers' voices to contest exactly such myths and although he ends over-pessimistically on the note that ultimately the media myth did swamp out the real democratic public debate that actually took place, the very existence of his book serves to contest that myth retrospectively. The key point he makes is that the effect of the whole Diana saga has been to 'negatively reposition' the royal family as 'Them' (47). The consequences of this unbalancing of the People's Monarchy paradigm have yet to be felt, but they will have huge consequences for British society as a whole and for M-O in particular. The need for M-O's founding vision of a transformed participatory mass society based on a dialectical combination of collective independence and individual agency is now greater than ever.

Conclusion: Mass-Observation Reassessed

The 'rediscovery' of M-O by Addison and Calder set in motion a process of changing our perception of the history of the Second World War and its consequences. In particular, it has helped to highlight how modes of representation developed by M-O, itself, privileged a universalised image of the working class as the embodiment of a Britishness, which still endures through powerful symbols such as the monarchy and the National Health Service. The historian Ross McKibbin has written a powerful description of the political and cultural shift caused by the war:

> By the end of the 1930s the Conservative Party had created a huge, heterogeneous, but stable coalition. There was nothing to suggest it was provisional; everything to suggest it was a natural historical outcome. The only obvious threats to it were external. In this sense the Second World War threw British history, and even more, English history, off course .... More or less everyone in the interwar years agreed that England was a democracy. The question was – whose democracy? Before the outbreak of war the question seemed to have been answered ... the ruling definition of democracy was individualist and its proponents chiefly a modernised middle class; in the 1940s the ruling definition was social-democratic and its proponents chiefly the organised working class. The class, therefore, which in the 1930s was the class of progress became in the 1940s the class of resistance (McKibbin 1998: 531–3).

However, as we have seen, a range of cultural and political thinkers including Orwell, Common and the M-O founders were convinced that if the individualist democracy of the 1930s could be culturally combined with a working-class collective identity, a classless society would come into existence. Such a process was identified in the new stratum of technically-minded workers, to which Jennings and Madge as media workers could be said to belong. The fate of this classless fraction was not the same as that of McKibbin's middle class, because wartime representations of national identity did not just privilege the working class but, more precisely, deployed a working-class imagery in order to symbolically reconcile modernity to a traditional order. It is only by reading M-O against itself in the manner that this book has adopted that the complex interrelationships of this process can be fully disentangled.

It is, therefore, not surprising that reception conditioned by the previously existing historical paradigms has often failed to register the significance of M-O. Perry Anderson argued that Britain had never produced a classical sociology, but only two displaced forms in the guise of social anthropology and literary criticism. M-O, formed from the influences of the anthropology of Malinowski and Bateson and the literary criticism of Richards and Empson, can be seen as exactly that missing sociology, albeit perhaps not so classical as Anderson would like. Instead, as Highmore suggests, M-O should be considered part of a wider tradition of continental avant-garde sociology following from Simmel's description of the ambiguous situation of the individual in the modern world. Like Simmel, M-O tried to develop an understanding of society from within by using surrealist and documentary techniques to render everyday understanding, which is simultaneously qualitative and quantitative, as a fully conscious process. As Chaney and Pickering argued: 'It was precisely because Mass Observation posed such radical questions about representing everyday life that their archives cannot now be pillaged indiscriminately by social historians looking for nuggets of information' (32–3).

In a more general sense, an emphasis on M-O as a primary source tends to relegate it to the category of social context, which as discussed, has enabled a particularly literary myth of the 'Thirties' to signify the incompatibility of literature and politics. This split between text and context can also be seen to contribute to the rise of cultural studies, which in the trajectory running through Eliot becomes negatively defined against the high modernist poetry that is privileged over it. The later development by cultural studies of its own branch of poetics suggests that poetic perception is somehow constitutive of consciousness so that, for example, life writing can be seen as a continuous act of existential affirmation in a context in which grand narratives are viewed with suspicion. Yet in the process the forms of cultural poetics and poetry have become sundered in such a way that a figure like Madge is doubly excluded from cultural centrality: he is both too poetic to be cultural and too cultural to be canonical. The reason that this is a problem is that life writing on its own is not a vehicle for social transformation. While it undoubtedly functions as a Foucauldian reverse discourse by cutting living space out of the materials provided by dominant institutions, it does not consistently produce images capable of transforming the symbolic order of society, as poetry can. Such claims have sometimes been considered elitist. For example, Swann has cited the opinion of wartime commercial exhibitors in order to suggest that Jennings's wartime films 'did not have the impact upon wartime cinema audiences that they had upon critics and film students in later years': 'symbolism may make its appeal to the few cultured minds, but propaganda to have its widest and strongest appeal, must speak what a former generation called "the vulgar tongue"' (Swann 1989: 165–6). Yet this is to miss the central anthropological point that symbolism affects all minds whether consciously or unconsciously. All cinema and advertising employs such means at some level, as do the newspaper headlines which Madge identified as forms of poetry. The point is not that one has to be 'cultured' to appreciate them or even necessarily resist them, but that one has to be aware how they work in order to create a different liberated order. This was what M-O tried to do in the 1930s by instructing their observers to record images.

The analytical division of poetic images by Richards into intellectual and active streams enabled Empson to formulate a form of ambiguous 'popular poetry' which allowed an image to be specifically interpreted even while indicating the possibility of its transformation into something else. This idea of forces held in a poised tension characterises modernist individual self-consciousness and the consequent possibility of agency. Jennings – amongst a host of other avant-garde film makers – recognised how technical advances in film could be used to expand the image space and the promise of modernity to the masses: the possibility of combining collective activity with individual agency into free identity. However, the problem was that the ownership of the necessary technology remained heavily concentrated: M-O could not compete with the media myth of the coronation of George VI. In this sense, Milne is correct in suggesting that 'the idea of Mass-Observation is perhaps more stimulating for rethinking the politics of the imagination than anything embodied by the actual results of Mass-Observation as an existing social practice' (69).

However, in a different sense, M-O did manage to influence the popular representation of British identity, to demonstrate what members of the public were thinking during the Munich Crisis and to change the tax structure of the country. Postwar Britain really did begin at home in 1938. Orwell comments in *The Road to Wigan Pier* that it is 'the memory of working-class interiors ... that reminds me that our age has not been altogether a bad one to live in' (102). It is the memory of the expansion of that culture across everyday life in a great pastoral pantomime of class that reminds us that the postwar age was similarly not altogether a bad one to live in. While M-O ultimately failed in the attempt to convert everyday life itself into a public site of contradiction and contestation on the model of politicised theatre, they certainly subverted the theatrical expression of public politics, which was so characteristic of the 1930s.

As Ben Highmore argues, M-O was precisely at its most productive and radically democratic when it blurred – as opposed to either upholding or dissolving – the distinction between 'native' informants and participant observers: 'It is here that Mass-Observation can be seen to fulfil the promise of Surrealist ethnography: the potential for everyone (academic ethnographers, capitalist industrialists, working men and women, and so on) to become "natives"' (Highmore 2002a: 87). It was in this sense that Harrisson argued that 'one had to be completely ordinary'. Yet this is only fun for those who do not have to remain 'ordinary', in the way that the influence of the media forces the current M-O diarists to describe themselves. M-O's description of the attraction of the Lambeth Walk was misleading in its suggestion that 'the working classes like to be Lambethians because Lambethians are like themselves'. The real pleasure came from a carnivalesque liberation similar to that of the coronation. However, the experience of writing up *Britain* must have been even more enjoyable for Madge because, as was the case with Jennings and the coronation, he was writing partly about his own experiences of participating in the carnival. Empson describes a scene from *The Beggar's Opera* in which a character is at 'the first level of comic primness and the author at the third' (Empson 1995:175). On this model, one can simultaneously enjoy being the 'native' and the 'ethnographer' and so combine the everyday pleasures of being ordinary with a philosophical independence which guarantees agency. The technical limitations which existed until very recently made this impossible at a mass level, but now it really is possible for new kinds of web-based M-O projects, allowing reports to be pooled and collectively edited into different configurations by all participants. In the future M-O, everyone will be both 'native' and 'ethnographer' and in possession of a poetic kind of thinking powerful enough to change reality in order to meet their collective mass social needs.

Bibliography and Sources

Abrams, Mark. *Social Surveys and Social Action*, London: Heinemann, 1951.

Adams, Mary. 'Memorandum on the Function of Home Intelligence', MAP: 1/A, 22 January 1940a.

—— 'Home Intelligence', MAP: 1/A, 16 February 1940b.

—— Memo to Macadam, MAP: 1/B, 20 February 1940c.

—— Letter to Harrisson, MAP: 2/E, 21 February 1940d.

—— Memo to Waterfield, MAP: 1/B, 5 March 1940e.

—— Memo to Leigh Ashton, MAP: 1/B, 8 March 1940f.

—— Memo to Macadam, MAP: 1/B, 8 March 1940g.

—— Memo to Waterfield, MAP: 4/E, INF 1/262, 12 March 1940h.

—— Memo to Ashton, MAP: 1/B, 1 April 1940i.

—— Memo to Ashton, MAP: 1/B, 16 April 1940j.

—— Memo to Sir Kenneth Clark, MAP: 1/B, 29 June 1940k.

—— Memo to Welch, MAP: 4/E, INF 1/262, 3 July 1940l.

—— Memo to Mr Charles, MAP: 1/B, 7 August 1940m.

—— Memo to DG, MAP: 1/B, 17 August 1940n.

—— Letter to Harrisson, MAP: 2/E, 2 September 1940o.

—— Memo to Macadam, MAP: 1/B, 23 September 1940p.

—— Memo to Macadam, MAP: 1/B, 26 September 1940q.

—— Letter to Harrisson, MAP: 2/E, 14 November 1940r.

—— Letter to Harrisson, MAP: 1/B, 19 November 1940s.

—— 'Notes on Present Position for the DG', MAP: 1/A, 7 February 1941a.

—— Letter to Stephen Spender, MAP: 2/E, 6 March 1941b.

—— Letter to Harold [Nicolson?], MAP: 2/E, 15 March 1941c.

—— Letter to Julian Huxley, MAP: 2/E, 15 March 1941d.

—— Letter to Harrisson, MAP: 2/E, 7 April 1941e.

—— Memo to Parker, MAP: 4/E, INF 1/262, 16 April 1941f.

Addison, Paul. *The Road to 1945: British Politics and the Second World War*, London: Quartet, 1977.

Anderson, Benedict. *Imagined Communities: Reflections on the Origin and Spread of Nationalism*, revised edition, London: Verso, 1991.

Anderson, Perry. 'Components of the National Culture' in *Student Power*, (eds) Alexander Cockburn and Robin Blackburn, Harmondsworth: Penguin, 1969.

Arrighi, Giovanni. *The Long Twentieth Century: Money, Power, and the Origins of Our Times*, London: Verso, 1994.

Auden, W.H. *The English Auden: Poems, Essays and Dramatic Writings*, ed. Edward Mendelson, London: Faber and Faber, 1986.

Barefoot, Brian. Unpublished Memoir, M-O A: Former M-O Personnel, Box 1, 1939.

—— Supplement Account, M-O A: Former M-O Personnel, Box 1, 1979.

Barker, Francis, Bernstein, Jay, Coombes, John, Hulme, Peter, Musselwhite, David and Stone, Jennifer (eds) *1936: The Sociology of Literature, Volume 2 – Practices of Literature and Politics*, University of Essex, 1979.

Bartlett, F.C. 'Suggestions for Research in Social Psychology' in Bartlett et al, 1939.

—— Ginsberg, M., Lindgren, E.J. and Thouless, R.H. (eds) *The Study of Society: Methods and Problems*, London: Kegan Paul, Trench, Trubner & Co.,1939.

Bateson, Gregory. *Naven*, second edition, Stanford, Calif.: Stanford University Press, 1958.

Benjamin, Walter. *Moscow Diary*, (ed.) Gary Smith, trs. Richard Sieburth, Cambridge, Mass.: Harvard University Press, 1986.

—— 'The Work of Art in the Age of Mechanical Reproduction' in *Illuminations*, trs. Harry Zohn, London: Fontana, 1992.

—— 'Fate and Character' in *Selected Writings, Volume 1: 1913–1926*, (eds) Marcus Bullock and Michael W. Jennings, Cambridge, Mass.: The Belknap Press of Harvard University Press, 1996.

—— '"An Outsider Attracts Attention" – on *The Salaried Masses* by S. Kracauer' in Kracauer 1998.

—— 'Surrealism' in *Selected Writings, Volume 2: 1927–1934*, (eds) Michael W. Jennings, Howard Eiland and Gary Smith, Cambridge, Mass.: The Belknap Press of Harvard University Press, 1999.

—— 'Eduard Fuchs, Collector and Historian' in *Selected Writings, Volume 3: 1935–1938*, (eds) Howard Eiland and Michael W. Jennings, Cambridge, Mass.: The Belknap Press of Harvard University Press, 2002.

Bennett, Tony and Watson, Diane (eds) *Understanding Everyday Life*, Oxford: Blackwell, 2002.

Benthall, Jonathan. 'A Pioneer who was nearly a Great British Eccentric', *Independent*, 22 February 2000.

Beresford, John. Note, MAP: 4/E, INF 1/329, 21 April 1939.

Berman, Marshall. *All That Is Solid Melts Into Air: The Experience of Modernity*, London: Verso, 1983.

Blunt, Anthony. 'Rationalist and Anti-Rationalist Art', *Left Review*, 2, 10 July 1936.

Borkenau, Franz. *The Totalitarian Enemy*, London: Faber and Faber, 1940.

Brendon, Piers. *The Dark Valley: A Panorama Of the 1930s*, London: Pimlico, 2001.

Breton, André. 'Limits not Frontiers of Surrealism' in Read 1936.

—— *Manifestoes of Surrealism*, trs. Richard Seaver and Helen R. Lane, University of Michigan: Ann Arbor, 1972.

Broder, Albert. 'The "Long Twentieth Century" in Economic History', http://econpapers.repec.org/paper/abphe1999/003.htm, 1999.

Bronowski, Jacob and Davies, Hugh Sykes (eds) 'Cambridge Experiment' in *Transition*, 19–20, June 1930.

Bulmer, Martin (ed.) *Essays on the History of British Sociological Research*, Cambridge: Cambridge University Press, 1985.

Burris, Val. 'The Discovery of the New Middle Class' in Vidich 1995.

Calder, Angus. 'Mass-Observation 1937–1949' in Bulmer 1985.

—— 'Introduction to the Cresset Library Edition' in Harrisson and Madge 1986.

—— *The People's War: Britain 1939–1945*, London: Pimlico, 1992a.

—— *The Myth of the Blitz*, London: Pimlico, 1992b.

—— Obituary: Charles Madge, *Independent*, 20 January 1996.

—— and Sheridan, Dorothy (eds) *Speak for Yourself: A Mass-Observation Anthology, 1937–49*, London: Jonathan Cape, 1984.

Calder-Marshall, Arthur et al. 'Why Not War Writers', Manifesto, in Orwell *CW* XIII, 2001.

Carey, John. *The Intellectuals and the Masses: Pride and Prejudice among the Literary Intelligentsia 1880–1939*, London: Faber and Faber, 1992.

Carswell, John. *Lives and Letters*, London: Faber and Faber, 1978.

Caudwell, Christopher. *Illusion and Reality*, London: Lawrence & Wishart, 1977.

Cawson, Frank. Interview with Angus Calder, M-O A: Former M-O Personnel, Box 1, 29 February 1980.

Chaney, David and Pickering, Michael. 'Authorship in Documentary: Sociology as an Art Form in Mass Observation' in Corner 1986.

Chapman, Dennis. 'Towards the Study of Human Ecology', *Pilot Papers*, 1, 1 January 1946.

—— Interview with Nick Stanley (2 reels), M-O A: Former M-O Personnel, Box 1, 23 February 1979a & b.

—— Letter, M-O A: Former M-O Personnel Box 1, 11 March 2002.

Common, Jack. 'Pease-Pudding Men' in *Revolt Against an 'Age of Plenty'*, Newcastle: Strongwords, 1980.

—— *Freedom of the Streets*, People's Publications, 1988.

Corner, John (ed.) *Documentary and the Mass Media*, London: Edward Arnold, 1986.

Cross, Gary (ed.) *Worktowners at Blackpool: Mass-Observation and Popular Leisure in the 1930s*, London: Routledge, 1990.

Crossman, Richard [Crux]. 'London Diary', *New Statesman*, 5 February 1971.

Cunningham, Valentine. *British Writers of the Thirties*, Oxford: Oxford University Press, 1988.

—— 'Marooned in the 30s', *Times Literary Supplement*, 19 August 1994.

Davenport, T.H.R. and Saunders, Christopher. *South Africa: A Modern History*, fifth edition, Basingstoke: Macmillan, 2000.

Eagleton, Terry. *Literary Theory: An Introduction*, Oxford: Blackwell, 1983.

Easthope, Anthony. 'Traditional Metre and the Poetry of the Thirties' in Barker et al 1979.

Eliot, T.S., 'Tradition and the Individual Talent' in *Selected Essays*, London: Faber and Faber, 1932.

—— *Notes Towards the Definition of Culture*, London: Faber and Faber, 1948.

—— *Collected Poems*, London: Faber and Faber, 2002.

Empson, William. *The Gathering Storm*, London: Faber and Faber, 1940.

—— *Seven Types of Ambiguity*, Harmondsworth: Penguin, 1961.

—— *Some Versions of Pastoral*, Harmondsworth: Penguin, 1995.

—— *The Complete Poems*, Harmondsworth: Penguin, 2001.

Finch, Janet. *Research and Policy: The Use of Qualitative Methods in Social and Educational Research*, Lewes: The Falmer Press, 1986.

Firth, Raymond. 'An Anthropologist's View of Mass-Observation', *Sociological Review*, XXXI, 2, April 1939.

Ford, Ford Madox. *England and the English*, Manchester: Carcanet, 2003.

Frazer, James. *The Golden Bough: A History of Myth and Religion*, London: Chancellor Press, 1994.

Freud, Sigmund. *Psychopathology of Everyday Life*, Harmondsworth: Penguin, 1938.

—— 'Remembering, Repeating and Working-Through' in, *The Standard Edition of the Complete Psychological Works of Sigmund Freud: Volume XII* (ed.) James Strachey, London: The Hogarth Press, 1958a.

—— 'Observations on Transference-Love', *SE XII*, 1958b.

—— 'On Narcissism', *The Pelican Freud Library 11: On Metapsychology*, (ed.) Angela Richards, Harmondsworth: Pelican, 1984a.

—— 'Beyond the Pleasure Principle', *PFL 11*, 1984b.

—— 'The "Uncanny"', *The Pelican Freud Library 14: Art and Literature*, (ed.) Albert Dickson, Harmondsworth: Pelican, 1985.

Fried, Albert and Elman, Richard M. (eds) *Charles Booth's London*, Harmondsworth: Penguin, 1971.

Garfield, Simon. *Our Hidden Lives: The Everyday Diaries of a Forgotten Britain 1945–1948*, London: Ebury Press, 2004.

Gascoyne, David. *Journal 1936–37*, London: Enitharmon Press, 1980.

—— *Collected Journals 1936–42*, London: Skoob Books, 1991.

—— *A Short Survey of Surrealism*, London: Enitharmon Press, 2000.

Gillmor, Dan. *We the Media: Grassroots Journalism by the People for the People*, Sebastopol, CA: O'Reilly, 2004.

Gloversmith, Frank (ed.) *Class, Culture and Social Change: A New View of the 1930s*, Brighton: Harvester Press, 1980.

Gordon, Harry. 'Notes on Mass-Observation in Bolton', M-O A: Former M-O Personnel, Box 2, 1980.

Gorky, Maxim, Radek, Karl, Bukharin, Nicolai and Zhdanov, Andrey. *Soviet Writers' Congress 1934*, London: Lawrence and Wishart, 1977.

Graves, Robert and Hodge, Alan. *The Long Weekend: A Social History of Great Britain 1918–1939*, London: Abacus, 1995.

Green, Timothy. *The Adventurers: Four profiles of contemporary travellers*, London: Michael Joseph, 1970.

Gurney, Peter. ' "Intersex" and "Dirty Girls": Mass-Observation and Working-Class Sexuality in England in the 1930s', *Journal of the History of Sexuality*, 8 (2), 1997.

Habermas, Jurgen. *The Structural Transformation of the Public Sphere*, trs. Thomas Burger with Frederick Lawrence, Cambridge: Polity Press, 1992.

Haffenden, John. 'Introduction' to Empson 2001.

Haffner, Sebastian. *Germany: Jekyll and Hyde*, trs. W. David, London: Secker and Warburg, 1940.

Hall, Stuart. 'The Social Eye of *Picture Post*' in *Working Papers in Cultural Studies 2*, Centre for Contemporary Cultural Studies, University of Birmingham, Spring 1972.

Hamilton, Ian. 'Against Oblivion', *Guardian*, 16 March 2002.

Harrisson, Tom. 'Birds of the Harrow District 1925–30', *The London Naturalist*, 1930.

—— *Letter to Oxford*, Wyck, Glos.: Reynold Bray, The Hate Press, 1933.

—— *Savage Civilisation*, Left Book Club Edition, London: Gollancz, 1937a.

—— 'Mass-Observation and the WEA', *The Highway*, 30, December 1937b.

—— 'Mass Opposition and Tom Harrisson', *Light and Dark*, II, 3, February 1938a.

—— 'Mass-Observation: A Reply', *New Statesman and Nation*, 12 March 1938b.

—— 'Whistle While You Work', *New Writing*, new series, I, Autumn 1938c.

—— Letter to Chapman, M-O A: M-O Former Personnel, Box 1 [under Chapman], November or December 1938d.

—— Memo accompanying letter to Adams, MAP: 4/F, 3 October 1939a.

—— Letter to Hilton, MAP: 4/E, INF 1/261, 2 November 1939b.

—— Letter to Adams, MAP: 4/F, 6 November 1939c.

—— Memo to Madge, M-O A: M-O Organisation and History, Box 1: CM – TH Correspondence, 18 January 1940a.

—— Memo to Madge, M-O A: M-O Organisation and History, Box 1: CM – TH Correspondence, 25 January 1940b.

—— 'Morale and the Future', *Us*, 2, 10 February 1940c.

—— Letter to Adams, MAP: 4/E, INF 1/262, 26 March 1940d.

—— Memo to MOI, MAP: 4/E, INF 1/262, 8 April 1940e.

—— 'Morale Now', M-O A: FR89, 30 April 1940f.

—— Article, *Us*, 15, 10 May 1940g.

—— 'How to Make a Better Britain', M-O A: FR156, 30 May 1940h.

—— 'Yardstick Memo', M-O A: FR250, June 1940i.

—— 'Public Opinion about Mr Chamberlain', M-O A: FR251, 5 July 1940j.

—— Article, *Us*, 18, M-O A: FR360, 1 August 1940k.

—— Letter to Adams, MAP: 4/E, INF 1/262, 5 September 1940l.

—— 'War Adjustment', *New Statesman and Nation*, 28 September 1940m.

—— Letter to Adams, MAP: 4/F, 15 March 1941a.

—— Letter to Taylor, MAP: 4/E, INF 1/262, 3 October 1941b.

—— Letter to Taylor, MAP: 4/E, INF 1/262, 9 October 1941c.

—— 'War Books', *Horizon*, IV, 24, December 1941d.

—— 'I'm Getting Fed Up with Spies', *Lilliput*, 10, 3, March 1942a.

—— 'Notes on Class Consciousness and Class Unconsciousness', *Sociological Review*, XXXV, 1942b.

—— 'Who'll Win?', *Political Quarterly*, XV, 1, January 1944.

—— 'Demob. Diary', *New Statesman and Nation*, 28 September 1946a.

—— Letter to Chapman, M-O A: M-O Former Personnel, Box 1 [under Chapman], 30 December 1946b.

—— 'The Future of Sociology', *Pilot Papers*, II, 1, March 1947a.

—— Letter to Chapman, M-O A: M-O Former Personnel, Box 1 [under Chapman], 24 March 1947b.

—— *World Within: A Borneo Story*, London: The Cresset Press, 1959.

—— (ed.) *Britain Revisited*, London: Gollancz, 1961.

—— 'Mass-Observation at Bromgrove', *New Statesman*, 11 June 1971.

—— 'Don't Know Where ...', *New Statesman*, 19 September 1975.

—— (ed.) *Living Through the Blitz*, London: Collins, 1976.

——, Jennings, Humphrey and Madge, Charles. 'Anthropology at Home', letter to the *New Statesman and Nation*, 30 January 1937.

—— and Madge, Charles. *War Begins At Home*, London: Chatto and Windus, 1940.

—— and —— *Britain by Mass-Observation*, London: The Cresset Library, 1986.

Harvey, J. Brian. 'Review of *Some Versions of Pastoral*', *Left Review*, 2, 5, February 1936.

Hayward, Ian. *Working-Class Fiction*, Plymouth: Northcote House, 1997.

Heimann, Judith. *The Most Offending Soul Alive: Tom Harrisson and His Remarkable Life*, Honolulu: University of Hawai'i Press, 1998.

Hey, V. *Patriarchy and Pub Culture*, London: Tavistock, 1986.

Highmore, Ben. *Everyday Life and Cultural Theory: An Introduction*, London: Routledge, 2002a.

—— (ed.) *The Everyday Life Reader*, London: Routledge, 2002b.

Hilton, John. Memo to Waterfield, MAP: 4/E, INF 1/261, 22 September 1939.

Hitchens, Peter. *The Abolition of Britain*, revised edition, London: Quartet, 2000.

Hobsbawm, Eric. *Age of Extremes: The Short Twentieth Century 1914–1991*, London: Michael Joseph, 1994.

Hood, Walter. Interview with Tom Harrisson, M-O A: Former M-O Personnel, Box 3, May 1972.

Howard, Anthony. *Crossman: The Pursuit of Power*, London: Jonathan Cape, 1990.

Howarth, Herbert, Thompson, Hector and Watkin, Bruce. Untitled Fragment, M-O A: TC 'Dreams', 'Dominant Images and Dreams 1937' 1/A.

Hubble, Nick. 'Charles Madge and Mass-Observation are At Home: From Anthropology to War, and After', *new formations*, 44, 2001.

—— *George Orwell and Mass-Observation: Mapping the Politics of Everyday Life in England 1936–1941*, D.Phil. Thesis, University of Sussex, 2002.

—— 'Decline of the English Penguin and Other Mass-Observations' in Conference Proceedings: *Approaches to Englishness*, Birmingham: Hart, 2003.

—— 'Imagined and Imaginary Whales: Benedict Anderson, Salman Rushdie and George Orwell' in *World Literature Written in English*, 40, 1, 2004.

Hynes, Samuel. *The Auden Generation: Literature and Politics in England in the 1930s*, Princeton University Press, 1982.

Jackson, Kevin (ed.) *The Humphrey Jennings Film Reader*, Manchester: Carcanet, 1993.

—— *Humphrey Jennings*, Basingstoke: Picador, 2004.

Jackson, T.A. 'Marxism: Pragmatism: Surrealism', *Left Review*, 2, 11, August 1936.

Jahoda, Marie. Review of *May the Twelfth*, *Sociological Review*, XXX, 2, April 1938.

Jeffery, Tom. *Mass-Observation: A Short History*, occasional papers, Centre for Contemporary Cultural Studies, University of Birmingham, 1978.

—— 'A Place in the Nation: The Lower Middle Class in England' in Koshar 1990.

Jennings, Humphrey. 'Review of *Surrealism*', *Contemporary Poetry and Prose*, 8, December 1936.

—— *Pandaemonium: The Coming of the Machine as Seen by Contemporary Observers*, (eds) Mary-Lou Jennings and Charles Madge, London: Picador, 1987.

—— and Madge, Charles. 'Poetic Description and Mass-Observation', *New Verse*, 24, February–March 1937a.

—— and —— (eds), *May the Twelfth: Mass-Observation Day-Surveys 1937*, London: Faber and Faber, 1937b.

—— and —— (eds), *May the Twelfth: Mass-Observation Day-Surveys 1937*, London: Faber and Faber, 1987.

Jennings, Mary-Lou (ed.) *Humphrey Jennings: Film-Maker, Painter, Poet*, London: British Film Institute, 1982.

Jones, Linda Lloyd. 'Fifty Years of Penguin Books' in *Fifty Penguin Years*, Harmondsworth: Penguin, 1985.

Jones, Peter (ed.) *Imagist Poetry*, Harmondsworth: Penguin Classics, 2001.

Kahn, Derek. Review of *The Road to Wigan Pier*, *Left Review*, 3, 3, April 1937.

Keating, Peter (ed.) *Into Unknown England 1866–1913: Selections from the Social Explorers*, Hammersmith: Fontana, 1976.

Keynes, John Maynard. Letter to Charles Madge, CMP: 21/5, 7 April 1941.

—— *Collected Writings XII: Economic Articles and Correspondence*, (ed.) Donald Moggridge, London: Macmillan, 1983.

—— *Collected Writings XXI: Activites 1931–39*, (ed.) Moggridge, London: Macmillan, 1982.

Khan, Naseem. 'Mysteries of the Aspidistra Cult', *New Statesman*, 13 February 1987.

Koshar, Rudy (ed.) *Splintered Classes: Politics and the Lower Middle Classes in Interwar Europe*, New York: Holmes & Meier, 1990.

Kracauer, Siegfried. *The Mass Ornament: Weimar Essays*, trs. Thomas Y. Levin, Cambridge, Mass.: Harvard University Press, 1995.

—— *The Salaried Masses: Duty and Distraction in Weimar Germany*, trs. Quintin Hoare, London: Verso, 1998.

Kuper, Adam. *Anthropology and Anthropologists: The Modern British School*, third edition, London: Routledge, 1996.

Laing, Stuart. 'Presenting "Things as They Are": John Sommerfield's *May Day* and Mass-Observation' in Gloversmith 1980.

Large, E.C. Review of *May the Twelfth*, *New English Weekly*, 30 December 1937.

Last, Nella. *Nella Last's War: A Mother's Diary 1939–45*, (eds) Richard Broad and Suzie Fleming, London: Sphere, 1983.

Lederer, Emil and Marschak, Jacob. 'The New Middle Class' in Vidich 1995.

Lefebvre, Henri. *Critique of Everyday Life*, Volume One, trs. John Moore, London: Verso, 1991.

Lewis, C. Day (ed.) *The Mind in Chains: Socialism and the Cultural Revolution*. London: Frederick Muller, 1937.

Light, Alison and Samuel, Raphael. 'Pantomimes of Class', *New Society*, 19/26 December 1986.

Lloyd, A.L. 'Surrealism and Revolutions', *Left Review*, 2, 16, January 1937.

Lukács, Georg. *History and Class Consciousness*, trs. Rodney Livingstone, London: Merlin, 1971.

Lynd, Robert S. and Lynd, Helen Merrell. *Middletown: A Study in American Culture*, New York: Harcourt Brace, 1929.

—— and —— *Middletown in Transition: A Study in Cultural Conflicts*, New York: Harcourt Brace, 1937.

MacClancy, Jeremy. 'Mass-Observation, Surrealism, Social Anthropology: A Present-Day Assessment', *new formations*, 44, 2001.

McKibbin, Ross. *Classes and Cultures: England 1918–1951*, Oxford: Oxford University Press, 1998.

McLaine, Ian. *Ministry of Morale: Home Front Morale and the Ministry of Information in World War II*, London: Allen & Unwin, 1979.

Madge Charles. Letter to Michael Roberts, CMP 21/1, December 1932.

—— 'Surrealism for the English', *New Verse*, 6, December 1933.

—— 'The Meaning of Surrealism', *New Verse*, 10, August 1934.

—— 'Review of *A Short Survey of Surrealism*', *New Verse*, 18, December 1935.

—— 'Review of *Problems of Soviet Literature*', *Left Review*, 12, 5, February 1936a.

—— 'Bourgeois News', *New Verse*, 19, February–March 1936b.

—— 'Popular Poetry', notes, CMP: Box 10, notebook (started 9 December 1934), 1936c.

—— Letter, *New Statesman and Nation*, 2 January 1937a.

—— 'Magic and Materialism', *Left Review*, 3, 1, February 1937b.

—— 'Oxford Collective Poem', *New Verse*, 25, April–May 1937c.

—— 'Comment on Auden', *New Verse*, 26–7, November 1937d.

—— 'Press, Radio and Social Consciousness' in Lewis 1937e.

—— *The Disappearing Castle*, London: Faber and Faber, 1937f.

—— Letter, *New Statesman and Nation*, 5 March 1938a.

—— Letter to Inez Spender, CMP: 8/3, 2 August 1938b.

—— Letter to Inez Spender, CMP: 8/3,19 August 1938c.

—— Letter to Inez Spender, CMP: 8/3,15 November 1938d.

—— 'Drinking in Bolton', *New Writing*, new series, I, Autumn 1938e.

—— Letter to Inez Spender, CMP: 8/3, 4 December 1938f.

—— 'Poetry, Time and Place', unpublished essay, CMP: Box 10, big blue notebook in envelope, 1938g.

—— Letter to Inez Spender, 16 January 1939a.

—— Letter to Keynes, CMP: 21/5, 18 April 1939b.

—— Memo to Harrisson, M-O A: M-O Organisation and History, Box 1: CM – TH Correspondence, 21 January 1940a.

—— Letter to Keynes, CMP: 21/5, 17 March 1940b.

—— Letter to Inez Spender, CMP: 8/3, 14 July 1940c.

—— Letter to Inez Spender, CMP: 8/3, 15 July 1940d.

—— Letter to Inez Spender, CMP: 8/3, 29 August 1940e.

—— 'War-time Saving and Spending – A District Survey', *Economic Journal*, IL, June–September 1940f.

—— 'The Propensity to Save in Blackburn and Bristol', *Economic Journal*, L, December 1940g.

—— 'Public Opinion and Paying for the War', *Economic Journal*, LI, April 1941a.

—— *The Father Found*, London: Faber and Faber, 1941b.

—— *War-Time Pattern of Saving and Spending*, Cambridge: Cambridge University Press, 1943a.

—— *Target for Tomorrow, No. 1: Industry After the War: Who is Going to Run it?*, London: The Pilot Press, 1943b.

—— (ed.) *Pilot Guide to the General Election*, London: The Pilot Press, 1945.

—— Editor's Commentary, *Pilot Papers*, 1, 4, November 1946.

—— Editor's Commentary, *Pilot Papers*, 2, 1, March 1947.

—— 'Reflections from Aston Park', *Architectural Review*, 104, 1948.

—— 'The Social Pattern of a New Town', *The Listener*, 17 February 1949.

—— 'Planning for People', *Town Planning Review*, XXI, 1950.

—— 'Tour of the New Towns', *New Statesman and Nation*, 8 September 1951.

—— Postscript to Harrisson 1961.

—— *Society in the Mind*, London: Faber and Faber, 1964.

—— 'The Birth of Mass-Observation', *Times Literary Supplement*, 5 November 1976.

—— Interview with Nick Stanley, M-O A: Former M-O Personnel, Box 4, 23 March 1978a.

—— Interview with Nick Stanley, M-O A: Former M-O Personnel, Box 4, 26 May 1978b.

—— Letter, *The Times*, 24 November 1978c.

—— Interview with Angus Calder, M-O A: Former M-O Personnel, Box 4, March 1979.

—— 'Thirties Revisited', Review of *Class, Culture and Social Change* (ed.) Gloversmith, *New Society*, 10 July 1980.

—— 'A Note on Images' in M.L. Jennings, 1982.

—— 'Autobiography', unpublished manuscript, CMP: Box 2/1, 1987.

—— *Of Love, Time and Places*, London: Anvil Press, 1994.

—— and Jennings, Humphrey. 'The Space of Former Heaven', *Life and Letters*, 13, 2, December 1935.

—— and —— 'They Speak for Themselves: Mass Observation and Social Narrative', *Life and Letters*, 17, 9, Autumn 1937.

—— and Harrisson, Tom. *Mass-Observation*, London: Frederick Muller, 1937.

—— and —— *First Year's Work*, London: Lindsay Drummond, 1938.

—— and —— *Britain By Mass-Observation*, Harmondsworth: Penguin, 1939.

—— and Barbara Weinberger, *Art Students Observed*, London: Faber and Faber, 1973.

—— and Peter Willmott, *Inner City Poverty in Paris and London*, London: Routledge and Kegan Paul, 1981.

Madge, John. *The Origins of Scientific Sociology*, London: Tavistock, 1963.

Malinowski, Bronislaw. 'A Nation-Wide Intelligence Service' in Madge and Harrisson, 1938.

Marshall, T.H. 'Is Mass-Observation Moonshine?', *The Highway*, 30, December 1937.

Marx, Karl. *Early Writings*, trs. Rodney Livingstone and Gregor Benton, Harmondsworth: Penguin, 1992.

Mazower, Mark. *Dark Continent: Europe's Twentieth Century*, Harmondsworth: Penguin, 1999.

Mengham, Rod. 'Bourgeois News: Humphrey Jennings and Charles Madge', *new formations*, 44, 2001.

Milne, Drew. 'Charles Madge: Political Perception and the Persistence of Poetry', *new formations*, 44, 2001.

Mitchison, Naomi. *Among You Taking Notes ...The Wartime Diary of Naomi Mitchison 1939–1945*, (ed.) Dorothy Sheridan, London: Phoenix, 2000.

M-O. 'A Thousand Mass-Observers', M-O A: FR A4, 1937a.

—— Draft Version of *First Year's Work*, M-O A: Organisation and History Box 1, Early Original Papers, 1937b.

—— 'Bolton Through the Ages', M-O A: Worktown Papers, Box 1, 1938a.

—— 'Social Factors in Economics', memo, M-O A: Worktown Papers, Box 1, 1938b.

—— *Us*, 13, 26 April 1940.

—— *People in Production*, Harmondsworth: Penguin, 1942.

—— 'A Report on Penguin World', FR2545A–C, December 1947a–c.

—— *War Factory*, London: The Cresset Library, 1987a.

—— *The Pub and the People*, London: The Cresset Library, 1987b.

Moggridge, Donald. *Maynard Keynes: An Economist's Biography*, London: Routledge, 1992.

MOI. 'Collecting Division Minutes', MAP: 4/E, INF 1/331, 18 May 1939a.

—— 'Home Publicity Division Minutes', MAP: 4/E, INF 1/316, 15 September 1939b.

—— Treasury to Waterfield, MAP: 4/E, INF 1/261, 23 September 1939c.

—— DDG [Waterfield] to Minister [Lord Macmillan], MAP: 4/E, INF 1/261, 27 September 1939d.

—— Marginal note by Macmillan, MAP: 4/E, INF 1/261, 30 September 1939e.

—— Waterfield to Macmillan, MAP: 4/E, INF 1/261, 10 October 1939f.

—— Macmillan to DDG, MAP: 4/E, INF 1/261, 11 October 1939g.

—— 'Report on the Work of Mass-Observation', MAP: 4/E, INF 1/261, 26 October 1939h.

—— Ashton to Harrisson, MAP: 4/E, INF 1/262, 9 April 1940a.

—— 'Home Intelligence: Notes of Decisions taken by the Director General at a Meeting held on Friday September 27th 1940', MAP: 1/A, 27 September 1940b.

—— Macadam to Adams, MAP: 4/E, INF 1/262, 30 September 1940c.

—— 'Office Circular No 73', MAP: 2/G, 9 April 1941a.

—— Welch to Harrisson, MAP: 4/E, INF 1/262, 1 September 1941b.

—— Taylor to Harrisson, MAP: 4/E, INF 1/262, 2 October 1941c.

—— MOI to M-O, MAP: 4/E, INF 1/262, 5 November 1941d.

Mortimer, Raymond. Review of *Britain*, *New Statesman and Nation*, 14 January 1939.

Mülder-Bach, Inka. 'Introduction' to Kracauer 1998.

Mulhern, Francis. *The Moment of 'Scrutiny'*, London: Verso, 1981.

—— *Culture/Metaculture*, London: Routledge, 2000.

Nairn, Tom. *The Enchanted Glass: Britain and Its Monarchy*, London: Radius, 1998.

Nicholls, Peter. *Modernisms: A Literary Guide*, London: Macmillan, 1995.

Nicholson, Basil. Introductory Note to M-O 1987b.

Oeser, O.A. 'Methods and Assumptions of Field Work in Social Psychology', *British Journal of Psychology*, XXVII, April 1937.

—— 'The Value of Team Work and Functional Penetration as Methods in Social Investigation' in Bartlett et al 1939.

Orwell, George. *The Road to Wigan Pier, Complete Works*, V, (ed.) Peter Davison, London: Secker & Warburg, 1998.

—— 'Review of *War Begins at* Home' in A *Patriot After All: 1940–1941, Complete Works*, XII, revised edition, (ed.) Peter Davison, London: Secker & Warburg, 2000a.

—— 'The Proletarian Writer', *CW* XII, 2000b.

—— *The Lion and the Unicorn*, *CW* XII, 2000c.

—— 'War-time Diary', *CW* XII, 2000d.

—— 'War-time Diary' in *All Propaganda is Lies: 1941–1942, Complete Works*, XIII, revised edition, (ed.) Peter Davison, London: Secker & Warburg, 2001a.

—— 'Review of *The Pub and the People* by Mass-Observation' in *Keeping Our Little Corner Clean: 1942–1943, Complete Works*, XIV, revised edition, (ed.) Peter Davison, London: Secker & Warburg, 2001b.

Outhwaite, William. *Understanding Social Life: The Method Called Verstehen*, second edition, Lewes: Jean Stroud, 1986.

Pear, T.H. 'Some Problems and Topics of Contemporary Social Psychology' in Bartlett et al 1939.

Plomer, William. Review of *May the Twelfth*, *The Listener*, Supplement, 13 October 1937.

Pocock, David. 'Afterword' in Jennings and Madge 1987.

Pyke, Geoffrey. Letter, *New Statesman and Nation*, 12 December 1936.

Raine, Kathleen. *Defending Ancient Springs*, London: Oxford University Press, 1967.

—— *The Land Unknown*, London: Hamish Hamilton, 1975.

Ray, Paul C. *The Surrealist Movement in England*, Ithaca: Cornell University Press, 1971.

Read, Alan. *Theatre and Everyday Life: An Ethics of Performance*, London: Routledge, 1993.

Read, Herbert (ed.) *Surrealism*, London: Faber and Faber, 1936.

Remy, Michel. *Surrealism in Britain*, Aldershot: Ashgate, 1999.

—— 'The Entrance of the Medium' in Gascoyne 2000.

Richards, I.A. *Science and Poetry*, London: Kegan Paul, 1926.

Richards, Jeffrey and Sheridan, Dorothy (eds) *Mass-Observation at the Movies*, London: Routledge & Kegan Paul, 1987.

Richardson, Maurice. Review of *May the Twelfth*, *Left Review*, 3, 10, November 1937.

Roberts, John. *The Art of Interruption: Realism, Photography and the Everyday*, Manchester: Manchester University Press, 1998.

—— 'Philosophising the Everyday: The Philosophy of Praxis and the Fate of Cultural Studies', *Radical Philosophy*, 98, November–December 1999.

Roberts, Michael (ed.) *New Signatures*, London: Hogarth, 1932.

—— (ed.) *New Country*, London: Hogarth, 1933.

Roughton, Roger. 'Surrealism and Communism', *Contemporary Poetry and Prose*, 4–5, August–September 1936.

Samuel, Raphael. *Theatres of Memory, Volume 1: Past and Present in Contemporary Culture*, London: Verso, 1996.

Sheridan, Dorothy. *Damned Anecdotes and Dangerous Confabulations: Mass-Observation as Life History*, Mass-Observation Archive Occasional Paper 7, Brighton: University of Sussex, 1996.

—— (ed.) *Wartime Women: A Mass-Observation Anthology*, London: Phoenix Press, 2000.

—— Street, Brian and Bloome, David. *Writing Ourselves: Mass-Observation and Literacy Practices*, Cresskill, NJ: Hampton Press, 2000.

Simmel, Georg. *On Individuality and Social Forms: Selected Writings*, (ed.) Donald N. Levine, University of Chicago Press, 1971.

Spender, Humphrey. *Worktown People: Photographs from Northern England 1937–38*, Bristol: Falling Wall Press, 1995.

Spender, Stephen. Letter, *New Statesman and Nation*, 19 March 1938.

—— Letter to Mary Adams, MAP 2/F, 17 March 1941.

Speier, Hans. 'Middle-Class Notions and Lower-Class Theory' in Vidich 1995.

Stanley, Liz. *The Archaeology of a 1930s Mass-Observation Project*, University of Manchester, Department of Sociology Occasional Paper No. 27, 1990.

Stanley, Nick. *The Extra Dimension: A Study and Assessment of the Methods Employed by Mass-Observation in its First Period 1937–40*, Ph.D Thesis (CNAA) 1981.

Stonier, G.W. Review of *May the Twelfth*, *New Statesman and Nation*, 9 October 1937.

—— 'Mass Observation and Literature', *New Statesman and Nation*, 26 February 1938.

Summerfield, Penny. 'Mass-Observation: Social Research or Social Movement?', *Journal of Contemporary History*, 20, 1985.

Surrealist Group in England, 'Declaration on Spain', *Contemporary Poetry and Prose*, 7, November 1936.

Sutherland, John. *Stephen Spender*, London: Viking, 2004.

Sussex, Elizabeth. *The Rise and Fall of British Documentary*, Berkeley: University of California Press, 1975.

Swann, Paul. *The British Documentary Film Movement, 1926–1946*, Cambridge: Cambridge University Press, 1989.

Symons, Julian. *The Thirties: A Dream Revolved*, revised edition, London: Faber and Faber, 1975.

Taylor, A.J.P. *The Origins of the Second World War*, Harmondsworth: Penguin, 1964.

—— *English History 1914–1945*, Harmondsworth: Penguin, 1970.

Thomas, James. *Diana's Mourning: A People's History*, Cardiff: University of Wales Press, 2002.

Thomas, Geoffrey. Diary, M-O A: Former M-O Personnel, Box 4b, 1939.

Trebitsch, Michel. 'Introduction' to Lefebvre 1991.

Trevelyan, Julian. *Indigo Days*, London: MacGibbon & Kee, 1957.

Trotter, David. *The English Novel in History 1895–1920*, London: Routledge, 1993.

Various, Review of *May the Twelfth*, *Life and Letters*, 17, 10, Winter 1937.

Vidich, Arthur J. (ed.) *The New Middle Classes: Lifestyles, Status Claims and Political Orientations*, Basingstoke: Macmillan, 1995.

Wagner, Gertrud. 'The Psychological Aspect of Saving and Spending' [M-O A: M-O Former Personnel, Box 4b], MA thesis, University of Liverpool, October 1939.

Willett, John. *The New Sobriety 1917–1933: Art and Politics in the Weimar Period*, London: Thames and Hudson, 1978.

Williams, Raymond. *The Country and the City*, London: The Hogarth Press, 1985.

—— *Culture and Society: Coleridge to Orwell*, London: The Hogarth Press, 1990.

—— *Orwell*, Hammersmith: Fontana, 1991.

Winston, Brian. *Claiming the Real: The Documentary Film Revisited*, London: BFI, 1995.

Wood, Kingsley. 'Financial Statement of the Chancellor of the Exchequer', *Hansard*, 7 April 1941.

Worsley, Peter. *The Trumpet Shall Sound*, London: Paladin, 1970.

Ziegler, Philip. *Crown and People*, London: Collins, 1978.

Index